Automated Machine Learning and Meta-Learning for Multimedia

Wenwu Zhu • Xin Wang

Automated Machine Learning and Meta-Learning for Multimedia

 Springer

Wenwu Zhu
Department of Computer Science
and Technology
Tsinghua University
Beijing, China

Xin Wang
Department of Computer Science
and Technology
Tsinghua University
Beijing, China

ISBN 978-3-030-88134-4   ISBN 978-3-030-88132-0   (eBook)
https://doi.org/10.1007/978-3-030-88132-0

This Springer imprint is published by the registered company Springer Nature Switzerland AG
The registered company address is: Gewerbestrasse 11, 6330 Cham, Switzerland

Foreword by Martin Ester

Machine learning (ML) has achieved spectacular breakthroughs in many real-life applications, including image classification, machine translation, and robotics. Deep neural networks have been particularly successful, due to their great capacity to approximate extremely complex, non-linear functions. Neural networks achieve this capacity through increasingly diverse and complex network architectures. However, the great potential comes at a hefty price: the development of machine learning models does not only require suitable training datasets and knowledge of the application domain but also deep knowledge of the machine learning methods to actually leverage their full potential.

Automated machine learning (AutoML) is a research direction that has recently emerged in response to the ever-increasing complexity of ML models and their development, aiming to automate the development process as much as possible. Goals are, in particular, to automatically tune the many hyperparameters, e.g., the number and size of layers, and to determine the most appropriate architecture of a neural network, e.g., a convolutional network for feature extraction combined with a fully connected network for classification. A related goal is meta-learning, i.e., learning to learn, which promises to reduce the effort of model development by transferring a model from a source domain to a target domain. While AutoML is still a fairly young research area, neural networks developed through AutoML have already achieved performance comparable to that of neural networks handcrafted by data scientists in some applications. Since 2015, the research community has organized the AutoML Challenge, which has provided a benchmark and much stimulation to the field. Several ML development tools, including RapidMiner and Microsoft Azure, have already implemented the features of AutoML. In conclusion, AutoML is a promising direction in ML that is expected to mature in the years to come.

This timely book by two experts in the field introduces the state-of-the-art in AutoML with a focus on it for multimedia data. Wenwu Zhu is a Professor at the Department of Computer Science at Tsinghua University and is widely recognized for his research in the areas of multimedia networking and computing as well as multimedia big data. Xin Wang is an Assistant Professor at the Department of

Computer Science at Tsinghua University. Multimedia data, including image, video, audio, and text data, is much more complex in nature than structured data such as records stored in a relational database, and multimedia has been the domain where deep neural networks have had the greatest impact. In addition to being unstructured, multimedia data is typically very large and multimodal, i.e., combines various types of multimedia data, e.g., images and text. Finally, multimedia data is often served in a streaming fashion, i.e., large amounts of data arrive rapidly and have to be processed online. I highly recommend this book to anyone who wants to understand the state-of-the-art in AutoML, in particular the special challenges and methods for multimedia data.

Professor, Simon Fraser University   Martin Ester
Burnaby, BC, Canada

Foreword by Steven C.H. Hoi

As a fundamental subset of AI techniques, machine learning has drawn popular attention from both academia and industry and made significant impact in real-world applications. This book covers two important and closely related subfields of machine learning, automated machine learning (AutoML) and meta-learning, which have been actively studied in recent years.

AutoML aims to automate the task of applying machine learning to solve a real-world problem. For example, one popular technique in AutoML is hyper-parameter optimization (HPO), which aims to automatically choose optimal hyperparameters for a learning algorithm. Another well-known AutoML technique is neural architecture search (NAS) for deep learning or deep neural network (DNN), which aims to automate the design of deep learning architectures.

Meta-learning, also known as learning to learn, aims to design a model that can learn new skills or adapt to new environments rapidly with limited training data. Meta-learning can be applied to tackle the AutoML tasks, such as HPO and NAS. In addition, meta-learning can be used for several other kinds of machine learning tasks and real-world applications, such as cold-start recommendation in multimedia and few-shot learning in computer vision and NLP.

This book provides a comprehensive understanding of AutoML and meta-learning methods and their applications. It is organized into two parts. Part I covers the subjects on the fundamentals of AutoML and meta-learning methodologies, including basics of some popular algorithms and recent advances in machine learning. Part II covers the subjects on applying AutoML and meta-learning techniques for a range of application domains, such as computer vision, natural language processing, multimedia, data mining, and recommender systems.

The authors are established AI experts and researchers with extensive experiences in investigating machine learning techniques for real-world applications. This book is strongly recommended for AI researchers, engineers, graduate students, and any readers who are interested in learning advanced machine learning subjects in AutoML and meta-learning.

Managing Director, Salesforce Research Asia   Steven C.H. Hoi
Professor, Singapore Management University
Singapore

Foreword by Tong Zhang

I have known professor Wenwu Zhu for many years, and we have collaborated on a number of problems. Professor Zhu is a highly regarded scientist in multimedia and big data research. He has not only published many influential scientific papers but also worked on practical problems in the industrial setting.

Both automated machine learning and meta-learning are emerging topics in machine learning, which have drawn significant attention in recent years due to their many practical applications. This book presents a comprehensive overview of the recent advances in these subjects as well as their applications. The book contains two parts. Part I presents a unified view of basic concepts and many recently proposed algorithms that are scattered in the literature. This helps the readers to quickly grasp the basic concepts and algorithmic foundations. Part II contains case studies and applications that help the readers to understand how these methods can be applied to real-world problems. Although examples in this book focused on multimedia applications, the material should greatly benefit all researchers and practitioners who want to learn these advanced machine learning methods through practical examples.

Chair Professor   Tong Zhang
The Hong Kong University of Science and Technology
Hong Kong, China

Preface

This book is to disseminate and promote the recent research progress and frontier development in AutoML and meta-learning as well as their applications in computer vision, natural language processing, multimedia, and data mining-related fields, which are exciting and fast-growing research directions in the general field of machine learning. We will advocate novel, high-quality research findings and innovative solutions to the challenging problems in AutoML and meta-learning. This topic is at the core of the scope of artificial intelligence and is attractive to audience from both academia and industry.

Our efforts in writing this book is motivated by the following reasons. First of all, the topics on meta-learning and AutoML are very new emerging topics, which urgently requires a well-organized monograph on these topics. Second, several current viewpoints may treat neural architecture search (NAS) and Bayesian optimization (BO), two important techniques in AutoML, as components in meta-learning. Our book differs from them by regarding AutoML and meta-learning as two parallel tools that can enhance each other. Third, in this book, we will discuss more recent advances in AutoML and meta-learning, such as continual learning, hardware-aware architecture search, and automated graph learning. Last but not the least, this book also focuses on the applications of AutoML and meta-learning in many research fields, such as computer vision, natural language processing, and multimedia etc.

Therefore, we deeply hope that this book can benefit interested readers from both academy and industry, covering the needs from junior starters in research to senior practitioners in IT companies.

Beijing, China   Wenwu Zhu

Beijing, China   Xin Wang
June, 2021

Acknowledgments

We thank Wenpeng Zhang for participating in the NeurIPS contest on AutoML in 2018, which initiated our research on AutoML and meta-learning. Many thanks to Jie Cai, Yudong Chen, Kun Cheng, Chaoyu Guan, Jiyan Jiang, Yuhang Jiang, Xiaohan Lan, Haoyang Li, Yue Liu, Yijian Qin, Guangyao Shen, Zheng Xiong, Yitian Yuan, Zeyang Zhang, Ziwei Zhang, and Shiji Zhou from the Tsinghua University for their contributions during the writing of this book.

We would also like to express our sincere thanks to Martin Ester from the Simon Fraser University, Steven C.H. Hoi from the Salesforce Asia/Singapore Management University, and Tong Zhang from The Hong Kong University of Science and Technology for writing forewords for this book.

Xin Wang is the corresponding author.

Contents

List of Figures

List of Tables

Part I
Basics and Advances for Automated Machine Learning and Meta-Learning

To make this book self-contained and cutting edge, **Part I** will cover both basic concepts and advanced research of AutoML and meta-learning.

Chapter 1
Automated Machine Learning

The last decade has witnessed a surge of machine learning (e.g., deep learning) research and applications in many real-world scenarios, such as computer vision, language processing and data mining. Most machine learning methods have a plethora of design choices that need to be made beforehand, and their performance is shown to be very sensitive to these choices. Furthermore, the desirable choices of algorithm design often vary over different tasks and hence the algorithm configuration requires intensive expertise, which becomes a substantial hurdle for new users and further restricts the applicability and feasibility of modern machine learning methods in a wider range of public fields.

To remedy this issue, automated machine learning (AutoML) is developed to configure machine learning methods in a data-driven, object-oriented and automatic way. AutoML aims to learn the configuration of machine learning methods that attains the best performance on the specific task. In this way, AutoML largely reduces the background knowledge needed to customize modern machine learning methods in specific application domains, which makes machine learning technologies more user-friendly.

In this chapter, we provide an overview of AutoML methods, including both fundamental principles and research frontiers. We first review the problem of hyper-parameter optimization (HPO), which gives the simplest and most general formulation of AutoML. We introduce three main branches of approaches to solve the HPO problem, namely, model-free methods, Bayesian optimization methods and bandit-based methods. For each individual branch, we discuss the basic concepts and principles behind it, and introduce representative methods. We shall note that, approaches in different branches are often integrated in the state-of-the-art HPO methods.

In the context of deep learning, the problem of AutoML becomes even more difficult, since modern large-scale neural networks often contain a vast amount of design choices, including neural architectures, regularization methods, training procedures, and all their respective hyper-parameters, and a single training or

W. Zhu, X. Wang, *Automated Machine Learning and Meta-Learning for Multimedia*, https://doi.org/10.1007/978-3-030-88132-0_1

evaluation procedure of a neural network is often time-consuming. Due to the complexity and significance of tuning neural networks, a new research field of AutoML is developed and successfully applied to deep learning scenarios, namely, neural architecture search (NAS). In this chapter, we also give a review of NAS. We discuss the principles behind NAS methods to remedy the issues of extremely large configuration space and very costly evaluation in automated deep learning, and introduce state-of-the-art methods proposed in this field.

In summary, this chapter gives a comprehensive overview of two main branches of automated machine learning, i.e., hyper-parameter optimization, which gives the most general problem formulation and learning strategies, and neural architecture search, which tackles the specific issues of enormous searching space and high evaluation expense in automated deep learning. Both principles and methods discussed here are generic, and can be tailored to specific task instances. In later chapters, we will see that these methods constitute the building blocks of state-of-the-art automated learning systems in the field of multimedia.

1.1   Hyper-Parameter Optimization

Modern machine learning systems are typically associated with a vast number of design choices. For example, when constructing a modern deep neural network [18], there are a large number of free parameters regarding the network architecture, regularization and optimization process etc. which should be taken into consideration. These design choices are normally decided before the learning process, and can significantly affect the effectiveness of the learning systems. However, in many real-world machine learning scenarios, manually finding a suitable configuration of all these free parameters are very difficult since these free parameters are often complex, high-dimensional and may have interactions with each other. The overwhelming complexity and heavy reliance on expert knowledge further prohibit the feasibility of modern machine learning systems in big data applications.

Hyper-parameter Optimization (HPO) aims to automatically select the optimal hyper-parameter configuration of a machine learning system/model given a learning task [29]. Instead of manually tuning each hyper-parameter based on expert domain knowledge, HPO decides these hyper-parameters combinatorially in a data-driven manner through an automated trial-and-error process. Compared to manual search, HPO benefits in following advantages.

- First, HPO largely reduces human efforts in designing various machine learning models. It is especially suitable for scenarios where the number of hyper-parameters is overwhelming, resulting in a search space which is too vast to explore manually. In this sense, HPO provides an automatic way to decide these hyper-parameters without any human interventions [25].
- Second, HPO is able to improve the performance of machine learning algorithms for different applications by tailoring the algorithms towards these applications

at hand. Indeed, HPO has been proved to provide better performance when compared with handcrafted configurations in many machine learning scenarios [36, 45].

- Third, HPO improves the reproducibility for scientific research, by offering a fair and open platform for comparing different methods with respect to specific learning problems [5].

The problem of HPO is formulated as below. To begin with, we introduce some necessary notations and definitions. Specifically, suppose a machine learning algorithm A has m tunable hyper-parameters $\lambda_1, \ldots, \lambda_m$, and each hyper-parameter λ_i may be selected from some certain domain Λ_i. Then the hyper-parameter space of the algorithm can be characterized as $\Lambda = \Lambda_1 \times \cdots \times \Lambda_m$, which is the Cartesian product of the domains of all hyper-parameters. Each combination of the hyper-parameters $\lambda \in \Lambda$ represents an instantiation (or configuration) of the algorithm. We further denote the algorithm preset with such configuration λ as A_λ.

It is worthy of pointing out that the range of Λ_i for any individual hyper-parameter λ_i is flexible, for instance, hyper-parameters can be real-valued (e.g., learning rate), integer-valued (e.g., number of layers) or categorical (e.g., the choice of an optimizer). The domain of each hyper-parameter is often bounded within a finite interval for practical issues [4]. We note that, different hyper-parameters may conditionally depend on each other, resulting in the actual configuration space to be a subset of the Cartesian product defined above. This is because choosing different values for some hyper-parameters may deactivate other hyper-parameters. For example, in a neural network, the number of nodes in the j-th layer is able to affect the model performance only if the number of layers in the neural network is no less than j. The Combined Algorithm Selection and hyper-parameter (CASH) problem [11, 49] can serve as another example where the choice of preprocessing schemes and machine learning algorithms is modeled as a categorical hyper-parameter which further determines the internal hyper-parameters of each scheme or algorithm.

HPO aims to select the best configuration λ of the algorithm A on a given learning task. More specifically, given a dataset $\mathcal{D}$ that characterizes the task, the objective of HPO is to search for the configuration λ^*, with which the algorithm attains the lowest validation loss, i.e.,

$$\lambda^* = \arg\min_{\lambda \in \Lambda} \mathcal{V}(A_\lambda, \mathcal{D}), \tag{1.1}$$

where $\mathcal{V}(A_\lambda, \mathcal{D})$ can be any validation protocol to evaluate the algorithm A_λ on the dataset $\mathcal{D}$ [6]. For example, it can be the K-fold cross-validation loss measured by any given loss function $\mathcal{L}$ [49] (e.g., the misclassification rate or the mean squared error). Concretely, we assume the dataset $\mathcal{D}$ is split into K cross-validation folds:

$$\mathcal{D} = \{(D_{train}^{(1)}, D_{val}^{(1)}), \ldots, (D_{train}^{(K)}, D_{val}^{(K)})\}, \tag{1.2}$$

where $k \in 1, 2, \cdots, K$ and each fold consists of non-intersecting training set $D_{train}^{(k)}$ and validation set $D_{val}^{(k)}$. Let $\mathcal{L}(A_\lambda, D_{train}, D_{val})$ measure the validation loss evaluated on validation set D_{val} using the model trained by algorithm A_λ on training set D_{train}, then the K-fold cross-validation protocol can be expressed as

$$\mathcal{V}(A_\lambda, \mathcal{D}) = \frac{1}{K} \sum_{i=1}^{K} \mathcal{L}(A_\lambda, D_{train}^{(i)}, D_{val}^{(i)}).$$

Besides cross-validation, there are also alternative validation protocols, including the hold-out method which splits the full dataset as $\mathcal{D} = (D_{train}, D_{val})$, the racing algorithm [37] which uses conservative concentration bounds to sift out models with bad performance. We refer interested readers to [6] for an overview of different validation protocols.

In the following, we will discuss three major branches of hyper-parameter optimization methods, i.e., model-free methods, Bayesian optimization methods and bandit-based methods.

1.1.1 Model-Free Methods

The simplest variants of model-free HPO methods are grid search and random search. In grid search, for each of the m hyper-parameter $\lambda_i, i = 1, \ldots, m$, a finite candidate set C_i is defined within the feasible domain Γ_i, which consists of the candidate values of such hyper-parameter λ_i for the given learning task at hand. Then each combination λ of hyper-parameters in the Cartesian product space $C_1 \times \cdots \times C_N$ is evaluated by training the machine learning model A_λ and then measuring its validation loss. The combination λ^* with the highest performance score (or the lowest loss) is selected as the most appropriate algorithm configuration. Despite its very simple implementation, grid search suffers from the curse of dimensionality as the number of candidate combinations grows exponentially with the number (i.e., m) of the hyper-parameters. Therefore, in practice, grid search is often adopted when the configuration space is not too large [39].

As an alternative model-free method, random search partially handles the curse of dimensionality by testing random combinations of hyper-parameters uniformly from the whole configuration space. As is recognized, random search works better than grid search in some scenarios when some hyper-parameters are more significant than others [3], as it can explore a wider range of such hyper-parameters than grid search. To clarify this point, consider the setting where we are allowed at most B times of evaluations. For each of the m hyper-parameters, we can only test $B^{1/m}$ different values when using grid search, whereas B different values of the hyper-parameter can essentially be evaluated by using random search. Besides, random search further benefits in the following two advantages over grid search.

1. Random search is more compatible with parallelization, since different candidate configurations can be sampled and evaluated completely independently.
2. Random search allows more flexible budget control, since it can stop at anytime when the budget for evaluation is exhausted, whereas the scheme of candidate configurations (which determines the total number of evaluations) in the grid search must be decided beforehand.

Random search is usually adopted as a baseline search strategy in HPO literature, since it is very simple to implement, and makes no assumptions on the algorithm being optimized or the learning task at hand, so that it is compatible with most general HPO problems. Additionally, the best configuration so far in random search will converge to the optimal configuration in expectation as the budget (or the total number of evaluations allowed) goes to infinity. In practice, random search is usually integrated with more complex model-based optimization methods to guarantee a minimal rate of convergence and encourage exploration [1]. Random search is also an effective approach to warm-start other more complex searching strategies, such as Bayesian optimization or bandit-based methods that will be discussed later. Specifically, random search is designed to uniformly explore the hyper-parameter space and often succeeds in finding configurations that enjoy acceptable validation performance, which can be then used as the initializations for other advanced search strategies. However, since random search does not introduce any heuristic property or underlying assumption on the algorithm as well as the learning task, it may take substantially longer time to specify a desirable configuration than the search strategies guided by some heuristics. In fact, the guided search strategies discussed below and in the following subsections will generally outperform grid search and random search [4, 32].

Population-based methods, including genetic algorithms and evolutionary algorithms, etc., are regarded as another important branch of model-free HPO methods [10, 43]. These methods typically maintain a population of configurations, and improve such population through iteratively deriving new generations with better configurations. Widely adopted operations to generate new configurations include the mutation (i.e., locally perturbing the values of some hyper-parameters in the given configuration to produce a new configuration), and the crossover (i.e., interleaving the values of hyper-parameters in two given configurations to produce a new one). Specifically, at each iteration, a population-based method takes one or more of the above operations over configurations within the old generation to derive new configurations, and discards the ones with relatively low performance upon evaluating these new configurations. One of the commonly used population-based methods is the CMA evolution strategy [16, 17], in which the configurations are sampled from a multivariate Gaussian whose mean and covariance are learned from historical configurations with high performance in each generation. In general, population-based methods are conceptually simple and able to tackle heterogeneous data types.

1.1.2   Bayesian Optimization

Bayesian optimization (BO) is a state-of-the-art framework for optimizing black-box functions that are expensive for evaluation [7, 42]. The problem of HPO can then be effectively solved via BO by treating the generalization performance (i.e., the validation loss $\mathcal{V}(A_\lambda, \mathcal{D})$) of models and hyper-parameter settings λ as a black-box function $f(\lambda)$. Such black-box function is often very expensive to evaluate, as for any certain configuration λ, we typically need to run a full training procedure of the model and evaluate its validation performance. In general, BO consists of a surrogate probabilistic model to be fitted towards the observations, as well as an acquisition function determining the optimal candidate hyper-parameters given the probabilistic model. At each iteration, the probabilistic model is trained to fit all historical observations of the black-box function. Then the acquisition function determines the utility of each candidate configuration, and chooses one configuration for the next evaluation, considering a trade-off between exploration and exploitation. The acquisition function is usually designed to be cheap to compute and often has close-form optimizers. In the following, we will introduce the general protocol of Bayesian optimization, then describe each component of BO individually.

A. Sequential Model-Based Optimization

In general, Bayesian optimization is developed upon the method of sequential model-based optimization (SMBO) [22, 23], which has been used in many applications where the evaluation of each candidate configuration is very expensive. Model-based algorithms approximate the black-box function f with a surrogate M that is relative cheaper to update and evaluate. At each inner loop, SMBO first generates a candidate configuration λ with the highest utility for evaluation given the current surrogate (which usually trades off exploration and exploitation), then evaluates such candidate. After observing its validation performance, such observation to used to update the surrogate, so that the surrogate model may approach the black-box function. The pseudo-code of SMBO is presented as follows (also see [4]).

B. Surrogate Models

Gaussian process is a traditional and one of the most commonly used probabilistic model as the surrogate of the black-box function [40]. It is very expressive and enjoys computational efficiency since it has a closed-form predictive distribution given any historical observations (i.e., the posterior probability distribution). Formally, a Gaussian process $\mathcal{G}_\lambda$ is characterized by a mean $m(\lambda)$ and a covariance function $k(\lambda, \lambda')$. In practice, the prior of the mean function $m(\lambda)$ is often assumed

---

Algorithm 1 Sequential model-based optimization (SMBO)

1: **Input:** Surrogate model M, acquisition function a.
2: **Initialize:** Observation history $\mathcal{H} \leftarrow \emptyset$, prior surrogate model M_1, and step index $t \leftarrow 1$.
3: **while** the resource is not exhausted **do**
4:     Generate the next candidate configuration $\lambda_t \leftarrow \arg\min_\lambda a(\lambda, M_t)$.
5:     Evaluate λ_t and observe its value $f(\lambda_t)$.
6:     Append $\mathcal{H} \leftarrow \mathcal{H} \cup (\lambda_t, f(\lambda_t))$.
7:     Set $t \leftarrow t + 1$.
8:     Fit a new model M_t to $\mathcal{H}$.
9: **end while**
10: **return** the best predicted configuration $\lambda^* \leftarrow \arg\min_\lambda a(\lambda, M_t)$.

---

to be constant. But surely, it can be nonconstant when we need to incorporate expert knowledge of the black-box function [42]. Now suppose we have previously observed the function values of some configurations, then for any given configuration λ, the predicted mean and variance of the black-box function at λ can be given in closed-forms, i.e.,

$$\mu(\lambda) = k_\lambda^\top K^{-1} y, \quad \sigma^2(\lambda) = k(\lambda, \lambda) + k_\lambda^\top K^{-1} k_\lambda, \tag{1.3}$$

where k_λ is the vector of covariances between λ and all previously examined configurations, K is the covariance matrix of all these evaluated configurations, and y is the observed values of these examined configurations.

There are several choices for the covariance function k (also termed the kernel function), which dictates the structure of the black-box functions we can fit. The Matern kernel is very flexible and commonly used family of kernels, which is parameterized by a smoothness parameter $\nu > 0$. For example, the most commonly used one is the Matern 5/2 kernel [45], which is defined as

$$k_{\text{Matern}5/2}(\lambda, \lambda') = \theta_0^2 \exp(-\sqrt{5}r)(1 + \sqrt{5}r + \frac{5}{3}r^2), \tag{1.4}$$

where $r^2 = (\lambda - \lambda')\Lambda(\lambda - \lambda')$ and $\Lambda = \text{diag}(\theta_1^2, \ldots, \theta_m^2)$ is a diagonal matrix. Such kernel is then parameterized by an amplitude θ_0 and m length scales $\theta_1, \ldots, \theta_m$, which can be learnt from historical observations $\mathcal{H}$.

Despite the much success of Gaussian processes in many scenarios of HPO, they have two main downsides. The first issue is that, its complexity scales cubically (i.e., $O(n^3)$) in the number n of historical observations, which is due to the inversion of the covariance matrix $K \in \mathbb{R}^{n \times n}$. In practice, to reduce its complexity, Cholesky decomposition can be adopted, which effectively drops the complexity to $O(n^2)$ [42]. However, such composition must be recomputed at each iteration (i.e., whenever the kernel parameters are changed). There are other works to reduce this computation cost via scalable approximations, such as sparse pseudo-input Gaussian processes (SPGPs), which approximate the full process by using a subset of the observation points (namely, m inducing pseudo-inputs where $m < n$) to estimate the kernel K [41, 44], as well as sparse spectrum Gaussian processes (SSGPs), which takes a similar approach to the spectral space of the kernel [31]. However, some previous works have concerned about the calibration of their uncertainty estimates and applicability to the standard setting of hyper-parameter optimization [35, 50].

Another problem of Guassian processes is their poor scalability to the dimension of hyper-parameters, as the total times of evaluations needed to realize the fixed resolution increases exponentially as the number of the hyper-parameters grows. Such issue limits their applicability to many real-world applications when the number of hyper-parameters is relatively large. To tackle this issue, [8] proposes a two-stage strategy for optimization and parameter selection for high-dimensional Gaussian processes. Some researches have also noted that, for some specific HPO problems, most dimensions have actually very little effect on the structure of the objective function, e.g., for HPO on neural networks [3] or solving some NP-hard problems [24]. Namely, these problems have actually low effective dimensionality, and hence the effectiveness of Gaussian processes is expected to be preserved.

Besides Gaussian processes, some other probabilistic models have also been applied to Bayesian optimization, which can be more scalable and flexible in some specific scenarios. Here we introduce three main kinds of alternative surrogate models. The first kind is the neural network, which is very scalable and expressive due to the flexibility in its construction. A naive way to apply neural networks to Bayesian optimization is to use them as a feature extractor and then use their output to build on Bayesian linear regression [46]. A more advanced approach is to use a Bayesian neural network trained with stochastic gradient Monte Carlo [47]. In general, neural networks appear to be faster than Gaussian processes in Bayesian optimization when the number of function evaluations are sufficiently large, which supports better scalability. The flexibility and expressiveness of deep neural networks can also enable to perform Bayesian optimization on more complex tasks.

Another type of surrogate models is the Random Forest. Compared to Gaussian processes, Random Forests are more suitable to handle larger, categorical and conditional hyper-parameter spaces [23]. In addition, Random Forests in Bayesian optimization generally enjoy better scalability, namely, their complexity for n observations can be at most $O(n \log n)$, much smaller than $O(n^3)$ or $O(n^2)$ when using Gaussian processes. Due to these benefits, in the prominent AutoML pipelines of Auto-WEKA [49] and Auto-sklearn [13], their Bayesian optimization modules are mainly built upon Random Forests as the surrogate model.

The final type of surrogate models we here introduce is the Tree Parzen Estimator (TPE) model [4, 5]. Instead of modeling the probability $p(y \mid \lambda)$ of evaluation score y on the configuration λ as in the aforementioned surrogate models, TPE estimates the density functions $p(\lambda \mid y < \alpha)$ and $p(\lambda \mid y \geq \alpha)$ respectively, where α is the percentile that distinguishes the "good" configurations from the "bad" ones. In a typical TPE model, 1D Parzen windows are often used to model the two distributions. Specifically, for any given configuration λ under examination, the ratio $p(\lambda \mid y < \alpha)/p(\lambda \mid y \geq \alpha)$ characterizes the utility of choosing such configuration, which is then used by the acquisition function to generate the candidate configuration for the next evaluation. In TPE, a tree of such Parzen estimators is used to handle conditional hyper-parameters and its effectiveness has been witnessed on many structured hyper-parameter optimization tasks [4, 5, 12]. Note that, it is also used as the building block of the Hyperopt-sklearn framework [30].

C. Acquisition Functions

The role of the acquisition function is to select the candidate configuration for the next evaluation, by leveraging the trade-off between exploration and exploitation. Commonly used acquisition functions include expected improvement (EI) [38], entropy search (ES) [19], predictive entropy search (PES) [20], upper confidence bound (UCB) [48], and so forth. We mainly introduce the first two types. Note that, [42] provides a detailed overview of the commonly used acquisition functions, and makes a comparison among them.

As the most commonly used acquisition function, expected improvement [38] measures the expected gain compared to the historical minimum, namely,

$$EI(\lambda, M) = \mathbb{E}_{y \sim p_M(y|\lambda)}[\max\{f_{min} - y, 0\}] = \int_{\mathbb{R}} \max\{f_{min} - y, 0\} p_M(y \mid \lambda) dy, \tag{1.5}$$

where $p_M(y \mid \lambda)$ is the probability distribution of the (stochastic) loss y w.r.t. the candidate configuration λ given the surrogate model M, and f_{min} is the minimum of the observed losses so far. In Gaussian processes, EI can be computed in a closed form, i.e.,

$$EI(\lambda, M) = (f_{min} - \mu(\lambda))\Phi(\frac{f_{min} - \mu(\lambda)}{\sigma}) + \sigma\phi(\frac{f_{min} - \mu(\lambda)}{\sigma}), \tag{1.6}$$

where ϕ and Φ represents the standard normal density and distribution function, and f_{min} is the best observation so far.

Another popular type of acquisition functions is entropy search [19], which selects candidate configurations based on the predicted information gain w.r.t. the optimum. Specifically, ES characterizes the probability distribution $p_{min}(\lambda \mid \mathcal{H}) = p(\lambda \in \arg\min_{\lambda} M_{\mathcal{H}}(\lambda))$, where $M_{\mathcal{H}}$ represents the surrogate model fitted to historical observations $\mathcal{H}$. The information gain at any configuration λ is then measured by the expected relative entropy (i.e., KL divergence) between the posterior probability $p_{min}(\cdot \mid \mathcal{H} \cup \{(\lambda, y)\})$ and the uniform distribution $u(\lambda)$, which is defined as

$$ES(\lambda, M) = \mathbb{E}_{y \sim p_M(y|\lambda)}[KL(p_{min}(\cdot \mid \mathcal{H} \cup \{(\lambda, y)\}); u(\lambda))]. \tag{1.7}$$

In practice, the calculation of p_{min} and the KL divergence can be conducted via several approximation techniques [19].

In general, there is no single acquisition function that outperforms other functions for all problem instances, namely, which acquisition function to choose may depend on the specific algorithm and learning problem being considered. Therefore, one may prefer to keep a portfolio of multiple acquisition functions to generate a group of candidate points at each iteration, and use a meta-acquisition function to choose the candidate point within this group for the next evaluation [21].

1.1.3 Bandit Variants

Bandit based algorithms or bandit variants, as another group of well-documented sequential modeling approaches, have also been employed to handle hyper-parameter optimization problem. Best arm identification [52], a well-researched problem in bandit area, aims to efficiently identify the best arm from a set of candidate arms with strong theoretical guarantees. Jamieson et al. [26] model the problem of hyper-parameter selection as non-stochastic best arm identification problem, where each arm corresponds to a fixed hyper-parameter setting. They propose an intuitive algorithm, Successive Halving algorithm (SH) (Algorithm 2) with a fixed budget for evaluation, which uniformly allocates the budget to a set of arms for a predefined number of iterations before discarding the worst half, then repeats the process until only one arm left.

SH runs s rounds and each round consumes $n_i \times r = \frac{B}{s}$ budget, thus the total budget will not exceed B. It throws out the worst half candidates in each round, and there only has one configuration left in the end, which is what SH believes the best candidate. SH is a simple but effective algorithm for selecting the best configuration, and has a tight bound of enough budget for identification as follows.

Theorem 1.1 *Assume that for all $t \in T$ with index $i \in [|T|]$, and the limitation exists $\lim_{r \to \infty} l_{t,r} = v_i < \infty$, where $v_1 < v_2 < \cdots < v_{|T|}$. There exists a non-*

Algorithm 2 Successive halving algorithm (SH)

Input: Budget B, a set T of n configuration candidates. $l_{t,r}$ denotes the loss of the configuration t with r budget
Initial: $s = \lceil \log_2(n) \rceil$, $n_0 = n, r = \frac{B}{ns}$
for $i = 0, 1, \ldots, s - 1$ **do**
  $n_i = \lfloor n_{i-1} 2^{-1} \rfloor, r_i = 2^i r$
  Run each candidate in T with r_i budget and evaluate the loss $l_{t,r_i}, t \in T$
  Throw out the worst half, and maintain a set T' of n_i candidates
  Let $T = T'$
end for
Output: Singleton element of T

---

increasing function of $t = r \geq 0$ *such that* $|l_{t,r} - v_i| \leq \gamma(r)$. *For any* $\epsilon > 0$
let

$$z_{SH} = 2 \left\lfloor \log_2(n) \right\rfloor (n + \sum_{i=1}^{n} \gamma^{-1}(\max\{\frac{\epsilon}{4}, \frac{v_i - v_1}{2}\})). \qquad (1.8)$$

If Algorithm 2 runs with budget $B \geq Z_{SH}$ *then the returned* $\hat{i}$ *with index* $\hat{i}$ *satisfies*
$v_{\hat{i}} - v_1 \leq \epsilon/2$.

The proof refers to Li et al.'s work [53]. From Theorem 1.1, if the SH algorithm runs with any budget $B \geq z_{SH}$, then a good enough configuration is guaranteed to be returned. This bound is smaller than non-adaptive uniform allocation, which allocates B/n to each the arms and picks the arm with the lowest loss. Therefore SH shows the advantage of adaptive throwing out than vanilla uniform method.

With the SH algorithm, a hyper-parameter optimizer can identify the best configuration from a set of candidates. However, HPO search space is usually too large or continuous, and thus SH needs to sample a discrete subset of the search space before running SH. Since v and γ are usually not known in practice, HPO agent is in a dilemma of setting the number of samples: if the iterative method converges very slowly for a given dataset (i.e., γ decrease slowly with r), or the randomly sampled candidates perform similarly well (i.e., $\{v_i\}$ are similar), then it would be better to sample fewer candidates thus each candidate has enough budget to converge and distinguish; if the iterative method converges very quickly for a given dataset (i.e., γ decreases quickly with r), or most of the randomly sampled candidates perform largely different (i.e., $\{v_i\}$ are largely different), then it would be better to sample more candidates for making the better use of the sufficient budget.

In [53], Hyperband (Algorithm 3) is proposed to trade-off between the number of candidate configuration samples and the input budget in large search spaces. It runs SH as the inner loop, where the outer loop iterates over different numbers of samples under the same budget.

Hyperband contains $s_{max} + 1$ SH modulars, each SH round uses $\left\lceil \frac{(s_{max}+1)R}{(s+1)} \right\rceil (s + 1) * (s+1) = B$ budget to identify from $n = \left\lceil \frac{(s_{max}+1)\eta^s}{(s+1)} \right\rceil$ configurations, subject to the constrain that at least one configuration is allocated R resources. Hyperband start with the most aggressive SH, which sets the max number of candidates to maximize the exploration. Then it reduces the samples by a factor of η until the final SH round, which is simple a classical random search where every configuration is allocated R resources. Intuitively, due to the adaptive selections of sample complexity, there exists one SH round that satisfies the condition of Theorem 1.1 such that the best candidate would be selected. By doing so, Hyperband addresses the trade-off dilemma as previous discussed in practice.

Bandit approach shows efficiency in finding a good configuration with less budget than bayesian optimization methods, thus be a better choice when dealing with un-precise HPO tasks. However, bandit-based HPO is not well in finding the best configuration in a continuous space as bayesian variants. Falkner et

---

Algorithm 3 Hyperband

Input: Budget B, search space H and throw out ratio η. l_t denotes loss from the configuration t

Initial: number of SH rounds $s_{\max} = \lfloor \log_\eta(R) \rfloor$, $R = B/(s_{\max}+1)$

for $s \in \{s_{\max}, s_{\max}-1, \ldots, 0\}$ **do**

   SH Initial: $n = \left\lceil \frac{(s_{\max}+1)\eta^s}{(s+1)} \right\rceil$, $r = R\eta^{-s}$

   Sample a set T of n configuration candidates from the search space H

   for $i = 0, 1, \ldots, s$ **do**

      $n_i = \lfloor n_{i-1}\eta^{-1} \rfloor$, $r_i = r\eta^i$

      Run each candidate in T with r_i budget and evaluate the loss $l_t, t \in T$

      Throw out the worst $1 - \frac{1}{\eta}$ ratio, and maintain a set T' of n_i candidates

      Let $T = T'$

   end for

end for

Output: Singleton element with smallest intermediate loss seen so far.

---

al. [12] propose a method named BOHB that combines bayesian optimization and Hyperband, thus taking advantage of both methods.

Hyperband determines how many configurations to evaluate with the adaptive budget allocation (i.e., the SH modular), by randomly sampling without utilizing the knowledge from previous rounds. BOHB replaces the random selection of configurations at the beginning of SH round, which applies Tree Parzen Estimator (TPE [4]) that is a Bayesian optimization approach. BOHB combines the precise sampling technique of Bayesian optimization, and applies the efficient identification method (i.e., SH). By doing so, BOHB has the advantages of both Bayesian optimization and bandit identification, thus finds the best configuration quickly and accurately.

1.2   Neural Architecture Search

Deep learning has made great success on a variety of fields, such as image recognition, object detection, and machine translation. This success is mainly due to its powerful capabilities of automatic feature extraction and representation. It has been proven that neural architecture design is crucial to the feature representation of data and the final performance, and much human efforts have been contributed to creating new architectures for certain tasks. However, the creation of neural architecture heavily relies on the researchers' prior knowledge and experience. Since designing an optimal architecture is labor intensive and time consuming, the idea of Neural Architecture Search (NAS), which aims to automatically design architectures, has attracted wide attention. There has been an insurgence in research efforts by the machine learning community that seeks proper NAS methods in multiple research areas during the past three years.

NAS methods have three main components, *search space*, *search strategy*, and *performance estimation strategy*. The search space defines which groups of architectures can be taken into consideration in principle. This definition influences the duration of search and the quality of the solution. However, search space also introduces human bias, limiting the chances to design an architecture beyond human knowledge.

Search strategy is the key part of NAS methods. It details how to explore the search space efficiently. Search strategy should make a good balance between exploring search space and tuning over the (good) candidate architectures. Most search strategies in NAS follow a search-evaluate feedback scheme, treating search strategy and performance estimation strategy as two detachable components. Search strategies model NAS problem in a black-box manner and provide suggestions for architectures in each running round of the search-evaluate procedure.

The performance estimation strategies are designed to evaluate performance of architectures suggested by search strategies in a black-box manner as well. The search strategy thus can continue to discover architectures with the best performance evaluated by the estimation strategy, without knowing the detailed evaluating process. Therefore, lots of classical optimization algorithms can be directly applied on NAS, such as Reinforcement Learning (RL), Evolutionary Algorithm (EA), and Bayesian Optimization (BO).

There are also NAS methods combining search and performance estimation strategy together in a joint optimization manner. For instance, differentiable architecture search (DARTS) [84] models neural architectures as learnable parameters, and the problem is formulated as a loss minimization task on the super network. The performance estimation is done with a mixture of various architectures on the super network and the loss is differentiable with respect to architecture parameters. Those architecture parameters are updated during the search process by gradient based methods. At the end of searching phase, the architecture parameters are discretized to form the optimal neural architecture. In this type of NAS methods (such as DARTS), the performance estimation strategy component cannot be easily replaced with others.

Performance estimation strategy affects the algorithm efficiency significantly, since the performance estimation process are required to repeat a large amount of times in the whole scheme. The simplest evaluation strategy is to directly train the architecture from scratch, which will cost an unaffordable amount of computational resources, although this evaluation is the most accurate. To improve efficiency, many other evaluation strategies such as insufficient training, weight sharing, and prediction are explored, which can save plenty of computational resources at the cost of introducing certain errors. It is obvious that the performance evaluated by most estimation strategies will not be perfectly accurate, thus NAS methods usually train the optimal architecture upon termination of searching process from scratch to get the accurate performance. The stage of searching for the optimal architecture is called searching phase, while the stage of training the optimal architecture from scratch is called evaluation phase.

1.2.1   Search Space

Search space defines which group of neural architectures NAS methods can discover in principle. Defining NAS search space not only influences the search efficiency, but also determines the best performance a NAS algorithm can achieve. From a computational perspective, neural networks can be represented as a set of transformations, and a neural network with k layers can be formulated as follows:

$$\mathbf{z}_k = o_k(I_k). \tag{1.9}$$

Here, $\mathbf{z}_k$ is the output of the k-th layer, o_k is the operation adopted at the layer, I_k is the input features of the k-th layer. The output of the whole neural network is the output of the last layer. Therefore, there are three things that should be determined by NAS methods:

1. the number of layers k
2. the operations at each layer
3. the input source of each layer

Firstly, the number of layers, i.e., k, is usually defined manually, and NAS methods are able to change k by removing a few layers. Secondly, the operation candidates are predefined according to the task. For example, in image classification task, operation candidates may include convolutions, pooling, concatenation, and addition. The operation candidates within different layers can be different, and for some layers the candidates can be even manually fixed. Thirdly, the input features can be either chosen from previous layers or the original raw data. Extra restrictions can also be added by human (e.g., the input of one particular layer should only be chosen from the last several layers). The search space in existing literature can be mainly grouped into two categories: the global space (or macro space) and the cell-base space (or micro space). The main difference between these two categories is their restrictions on the input feature choices. We remark that the operation candidates tend to vary for tasks in different areas.

A. Global Space

In global search space, NAS method is expected to search all necessary components of the neural architecture. Therefore, global search space tends to be relatively large. A typical global search space is *chain-structured search space*, which can be formulated as follows:

$$\mathbf{z}_k = o_k(\mathbf{z}_{k-1}). \tag{1.10}$$

In this search space, each layer only receives its input from the last layer. As such, NAS methods only need to focus on searching for the number of layers and the

Fig. 1.1 Global search space: chain-structured search space [54]

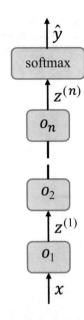

operations, without caring too much about searching for any input sources. Baker et al. [54] adopt such a kind of search space, as shown in Fig. 1.1.[1] They consider a set of operation candidates including convolutions, pooling, linear transformations with activation, and global average pooling, etc. The hyper-parameters in these operations, such as number of filters, kernel size, stride, and pooling size, are also taken into account. The process of searching the number of layers is controlled by the NAS method as well. To improve effectiveness and efficiency, some human prior knowledge is used to discard certain architectures with potentially poor performances or expensively computational costs. For instance, architectures with pooling as the first operation are removed from the search space. Furthermore, architectures with dense layers as high-resolution feature maps, or as feature transformations before other operations like convolutions are excluded from their search space.

In a parallel work around the same time, Zoph and Le [123] improve the chain-structured search space by adding skip connections, introducing input resource search into global search space. The space is shown in Fig. 1.2. For each layer, NAS methods should determine whether a previous anchor point is connected to current layer. Therefore, layers can receive inputs from previous layers other than the last one. If so, the input resource is also determined, and a concatenation operation is added to combine these inputs together. This design include a wider variety of architectures in the search space instead of searching for a single chain. However,

---

[1] This figure is from [107].

Fig. 1.2 Global search
space: chain-structured space
with skip-connection [123]

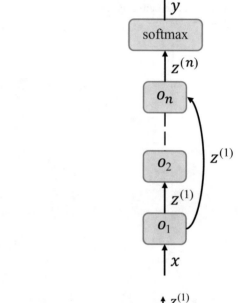

Fig. 1.3 Global search
space: search space with
segments [108]

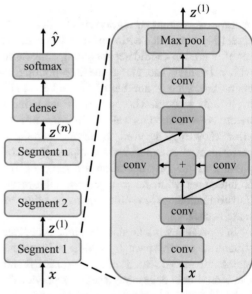

the operation candidates in this work is restricted to convolutions with different
hyper-parameter settings.

Xie and Yulie [108] construct a global search space in another way. They divide
the whole architecture as sequentially connected segments. Each segment contains
several layers, as shown in Fig. 1.3. For each layer, inputs are only chosen from
layers in the same segment, limiting the width of search space. This type of search

space is not a chain-structured search space because layers may not receive the input from the last layer. The authors also add some prior restrictions. The layers in segments are all convolutions. In addition, as a fixed part, each segment begins with a convolution layer and concludes with a max pooling layer with a stride of two to reduce feature dimensions. Therefore, this type of search space mainly focus on discovering the node connections in the architecture. The number of layers in each segment is predefined, but the NAS method can drop some layers by giving no connections between them and other layers, enabling NAS to tune the number of layers in a certain degree.

B. Cell-Based Space

Since NAS methods have heavy burden searching in global search space, a lighter search space is desired. Observing that many effective handcrafted architectures are designed with repetitions of small fixed parts, cell-based space is created. Those small fixed parts are called *cells*, which contains a relatively small number of operations. Cells are considered capable of information extraction. To construct the whole architecture, a few types of cells are usually stacked to get deeper and enhance the representation ability of the architecture. Since the inside structure of cells are the same, NAS methods only need to design a small structure, which greatly reduces the search space while constructing architectures with similar sizes. While NAS methods only search for the inside structure of cells in this space, how cells construct the whole architecture for different tasks is predefined, which needs expert knowledge or other algorithms. Another important advantage brought by cell design is flexibility. Different datasets and tasks may need architectures with different representation capabilities. When a few cells is designed on one task, it can be conveniently transferred to other tasks by changing the number of stacked cells and construction pattern of different cells.

NASNet [124] explores cell-based space in image tasks for the first time. The search space is as shown in Figs. 1.4 and 1.5.[2] It proposes to search for two types of cells, i.e., normal cells and reduction cells. Each cell has b *blocks*, where b is a hyper-parameter (with $b = 5$ being the most popular choice). Blocks are smaller structure units inside cells. Each block has two inputs and their corresponding unary operation with an addition at the end, which can be formulated as:

$$\mathbf{z}_k = o_k^{(1)}(I_k^{(1)}) + o_k^{(2)}(I_k^{(2)}). \tag{1.11}$$

Blocks' inputs can be chosen from blocks in the same cell, or results of the last two cells. Thus, unary operations and addition appear in turns. This scheme somehow restrict the chances to generate more various architectures. The output of a cell is defined as concatenation of the outputs of all blocks within the cell that do not serve

---

[2] These figures are taken from [124].

Fig. 1.4 Cell-based search
space in NASNet [124]:
whole architecture
constructed by cells

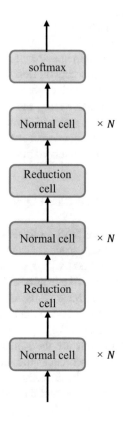

as inputs to any other block. The spatial resolution is kept unchanged in normal cells
while initial operations in reduction cells have a stride of two to reduce the spatial
resolution. Several normal cells are stacked and followed by a reduction cell to form
a stage. This stage is repeated for multiple times to construct the whole architecture.
To further improve searching efficiency, we can construct a proxy network with
fewer layers in search phase. The evaluation performance of proxy network reflects
the cells' effectiveness as well. When the optimal cells are designed, bigger network
is constructed for the evaluation phase.

Subsequent NAS works follow the cell-based search space with slight changes.
BlockQNN [121] does not use the block structure in cells, which is shown in
Figs. 1.6 and 1.7.[3] It treats unary operations like convolution and pooling equally
with binary operations such as concatenation and addition. Each operation in cells
can be any of these operation candidates, and the number of inputs depends on its
demand. Therefore, this alternative space definition has more freedom to generate
wider variety of the cell structures. Another difference is BlockQNN only search for
one type of cell, the reduction part is fixed as a max pooling layer.

---

[3] These figures are taken from [121].

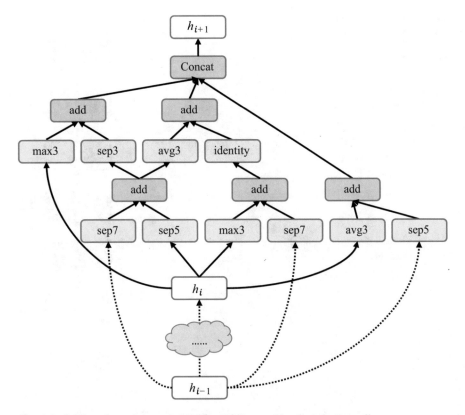

Fig. 1.5 Cell-based search space in NASNet [124]: a sample of a reduction cell

Liu et al. [83] propose a similar search space by hierarchical scheme. They use motifs to represent structure with different sizes. A high-level motif is composed several low-level motifs. level-1 motifs consist of basic operations such as convolution and pooling. Motifs construct a higher level motif by edges. A motif merges all inputs from the incoming edges by concatenation followed by a 1×1 convolution to keep the number of dimensions unchanged. A normal cell is a level-3 motif in their definition. The reduction part is fixed as a convolution with a stride of two.

Besides, FPNAS [67] considers the diversity of cells. It tries to stack diversified cells and finds the diversity benefits neural architecture performance. Designing more types of cells increases the burden of NAS methods, but creates chances to generate more complex architectures. We can even use more diversified cells to construct the architecture, which actually expands the search space. If we individually search diversified cells for all positions, the search space becomes global space, raising the upper-performance while reducing search efficiency. Therefore, it should be balanced to determine the size of search space and efficiency.

Xie et al. [109] study the performance of random architectures. They find that even architectures with random connection topology can achieve comparable per-

Fig. 1.6 Cell-based search
space in BlockQNN [121]:
whole architecture
constructed by cells

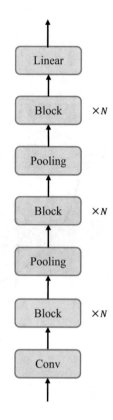

formances with expert-designed architectures, showing that the designing of search space already contains prior knowledge. Shu et al. [100] study the commonality of the cells obtained from popular NAS methods [84, 93, 95, 110, 124]. They find architectures tend to favor wide and shallow cells, showing this type of architectures are easier to converge during process both empirically and theoretically. However, NAS methods suffer from poor generalization performance. Since the search space significantly influences the performance, it is better to compare different search strategies in the same search space to guarantee fairness. Newly proposed search space should be evaluated by random search to indicate the performance gain by NAS methods.

1.2.2   Search Strategy

Search strategy controls the exploration in the search space. To accurately understand NAS problem, we give the formal definition of the problem here following [57, 107]. Firstly, a mapping Λ is defined as follows:

Fig. 1.7 Cell-based search space in BlockQNN [121]: a sample of a cell

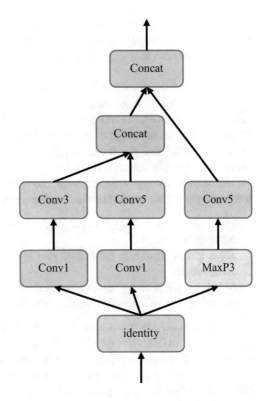

$$\Lambda : D \times A \to M. \tag{1.12}$$

Here, D denotes the space of all datasets, A denotes the architecture search space, and M denotes the space of all deep learning models. Given a dataset d, which is split into a training partition d_{train} and a validation partition d_{valid}, the general deep learning algorithm Λ estimates the model $m_{\alpha,\theta} \in M_\alpha$, where α denotes the architecture and θ denotes the weight. The model is estimated by minimizing the sum of a loss function $\mathcal{L}$ and a regularization term $\mathcal{R}$ with respect to the training data. That is,

$$\Lambda(\alpha, d) = \underset{m_{\alpha,\theta} \in M_\alpha}{\arg\min} \ \mathcal{L}(m_{\alpha,\theta}, d_{train}) + \mathcal{R}(\theta). \tag{1.13}$$

Given a dataset d, the neural architecture search task aims to finding the architecture α^* which maximizes an objective function O on the validation data d_{valid}. Formally,

$$NAS(d) = \alpha^* = \underset{\alpha \in A}{\arg\max} \ O(\Lambda(\alpha, d_{train}), d_{valid}) = \underset{\alpha \in A}{\arg\max} \ f(\alpha). \tag{1.14}$$

The objective function O can be the same as the negative loss function $\mathcal{L}$. The function f indicates the performance of architecture α. For the classification tasks, it is often the case that $\mathcal{L}$ is the negative cross-entropy and O is the classification accuracy.

As aforementioned, most NAS methods follow a feedback scheme, where the performance $f(\alpha)$ is modeled as a black-box. The calculation of f is given by performance estimation strategy which can be detached with the search strategy. There are also some strategies that should be jointly used with certain performance estimation strategies. For those search strategies, we will detail their performance estimation strategies along with them.

The search strategy is aiming to find the optimal architecture which has the best feedback $f(\alpha)$. The whole feedback scheme can be described as following steps.

- First, the search strategy gives architecture suggestions according to performance estimation history.
- Then, the performance estimation part gives the performances of the suggested architectures.
- At last, search strategy learns from the performances and updates its own model, preparing to give suggestions in the next turn.

There are two key parts that should be designed in search strategy, i.e., how to model the architecture space to efficiently explore in it and how to update the model. Classical classification of search strategies is based on the latter one, e.g., methods updated by Reinforcement Learning (RL) or Evolutionary Algorithm (EA). However, we should also pay attention to the modeling approach to the search space, which contains prior assumption about the architecture space, and guides NAS methods on how to explore the whole space. For instance, Recurrent Neural Network (RNN) is often used to model the distribution of architectures, which is under the assumption that layers in an architecture is sequentially determined. In this chapter, we follow the classical classification methods and divide search strategy into the five categories below.

A. Full Search and Random Search

Full search is the simplest approach to the optimization problem. Since NAS space is discrete, the number of possible architectures is finite. In condition that there are enough computational resources, we can estimate performances of all possible architectures, and the one with maximum performance is the optimal architecture. This approach theoretically achieves the best performance. However, there are too many possible architectures in the search space, and training all those architectures to get accurate performances is impossible with current computational resources. Therefore, full search is often combined with an extremely efficient performance estimation strategy. In practice, prediction-based NAS methods [62] choose full search to maximize their performance. Those methods train a predictor at first, then use the model to predict the performances of all architectures in search space. The computational burden of prediction is not heavy, and thus full search is acceptable.

Random search is another trivial method, which has no model updating process and the architecture space is modeled as uniform distribution. A significant advantage of random search is its flexibility. A certain number of architectures are randomly sampled from the search space according to the computational budget. Those architectures are evaluated and the one with the best performance is chosen. Xie et al. [109] run random search method on a global search space, where an architecture can be represented as a Direct Acyclic Graph (DAG). The nodes denote operations and the edges denote data flows. Nodes aggregate input data by a weighted sum, where the weights are learnable and positive. Then the data is processed by a transformation defined as a ReLU-convolution-BN triplet, i.e., a ReLU function, a convolution layer and a batch normalization layer are sequentially performed. The same copy of the transformed data is then distributed by the output edges. To construct the DAG, several classical random graph models are used, including Erdos-Renyi, Barabasi-Albert, and Watts-Strogatz. The authors find that those architectures with random connection topology can achieve comparable performances with expert-designed architectures on image classification and object detection tasks.

In addition, random search is also used as a baseline to indicate the performance gain of NAS methods [84]. The performance of random search reflects the performance level of the search space. Although it is shown that random architectures can perform well, the absolute performance cannot prove the effectiveness of NAS methods in a certain search space. It is more convincing to gain a large margin above performance of random search for other NAS methods.

B. Reinforcement Learning

Reinforcement Learning (RL) is a powerful approach to sequential decision problems. RL is modeled as the interaction of an agent and the environment. The purpose of RL is to teach agent to make good decisions based on the feedback of the environment. States S and actions $\mathcal{A}$ are important concepts in RL, which is predefined according to the task. States describe the condition of the agent, and actions are decision candidates of the agent. At each step t, the agent is at a state s_t, and it should choose an action a_t by a policy $\pi : s \rightarrow a$ to execute. Then the environment will transit the agent to another state s_{t+1}, seeding a reward r_t as feedback at the same time. RL optimizes the policy π to get the maximum discounted reward $\sum_0^T \lambda r_t$, where λ is the discount factor.

Solving the NAS problem with RL is modeling the NAS as a sequential decision problem. There are two main approaches to define the decision making process.

- Decide the architecture layer by layer. Later layers are determined based on all the previous ones, and the already decided layers cannot be changed again.

- Decide how to improve the architecture once at a time with the current information about the whole architecture. Any layer may be modified at any step. Examples of such modifications include widening and operation changing.

Since the sequential decision process starts from initial state, and different actions are taken at each step, the architecture space is modeled as a tree-structured space. The action distributions of all tree nodes (states) are learned by RL according to the feedback performances.

Q-Learning Methods

There are mainly two types of methods to solve a RL problem, i.e., the value-based methods and the policy-based methods. The value-based methods use V-value and Q-value to model the expectation of total discounted reward after each state and action. They can be formulated as:

$$V(s_t) = \mathbb{E}_\pi(a|s_t)[\mathbb{E}[r_t|s_t, a] + \gamma \mathbb{E}_{s_{t+1}|s_t, a} V(s_{t+i})]. \tag{1.15}$$

$$Q(s_t, a) = \mathbb{E}[r_t|s_t, a] + \gamma \mathbb{E}_{s_{t+1}|s_t, a} Q(s_{t+1}, a). \tag{1.16}$$

The V-value can be represented by Q-value:

$$V(s_t) = \mathbb{E}_\pi(a|s_t)[Q(s_t, a)]. \tag{1.17}$$

Therefore, if we obtain accurate Q-value of all states and all actions, we can conduct a policy π that chooses the action with the maximal Q-value to maximize the discounted reward of the current state, solving the RL problem. The optimization problem is modeled as Bellman Equation:

$$Q(s_t, a) = \mathbb{E}_{s_{t+1}|s_t, a}[\mathbb{E}[r|s_t, a] + \gamma \max_{a' \in \mathcal{A}(s_{t+1})} Q(s_{t+1}, a')]. \tag{1.18}$$

However, Bellman Equation is hard to solve directly, and therefore a series of functions Q_i are used to approximate the real Q-function Q in practice, with Q_i being updated in an iterative way:

$$Q_{i+1}(s_t, a) = (1 - \alpha)Q_i(s_t, a) + \alpha[r_t + \gamma \max_{a' \in \mathcal{A}(s_{t+1})} Q_i(s_t, a')], \tag{1.19}$$

where α is the learning rate to control the iteration speed. As states and actions are both discrete, traditional Q-learning algorithm uses an array in memory to record Q-values. We just refer to the array to acquire for a Q-value. Similarly, Q-values are updated by changing the values in the array.

Baker et al. [54] make one of the first attempt to propose to utilize Q-learning to solve NAS problem on image tasks. They use a global chain-structured search space, and the decision making process is to decide the operations in the chain one by one. The state is defined as a tuple of layer parameters, including the operation type of the last layer, layer depth, and representation size. The layer depth denote how many layers have been determined. The representation size denote the spatial resolution of the image after previous operations. The operation types include convolution, pooling, fully connected layer, global average pooling, and softmax. Those operations also have their own hyper-parameters. As for the action space, the agent can only construct one layer by one action, i.e., to add the layer depth from i to $i + 1$, guaranteeing sequential construction of the architecture. Prior limitations on operation selection should also be observed to exclude some expensively computational processes. For example, consecutive pooling layers are not permitted, fully connected layers can only be adopted at the last two layers, and the representation size must be small. The representation size is calculated according to the current representation size and the stride of selected operation. Besides, the agent can terminate at any state. There is also an upper bound to restrict the maximal layers of the architecture.

In Q-learning process, the discount factor is set as $\gamma = 1$, i.e., the authors do not care about the number of layers, but only the final performance. ϵ-greedy algorithm and experience replay are contained in their scheme, where ϵ is the hyper-parameter to balance the exploration and exploitation. At the beginning, the agent uses a large ϵ to explore a wide space. Then ϵ decreases to exploit existing knowledge more, further optimizing the performances. The agent continues to take action according to ϵ and the current policy until terminating to construct architectures. Finally, these architectures are evaluated and the validation accuracy is the feedback reward. In the other states except for the terminating state, no rewards will be given. The trajectory and the corresponding reward are combined and saved into experience. The Q-values are updated by replaying those experience and construct a new policy for next episode. The optimal architecture is given by the optimized policy after the last episode.

Zhong et al. [121] also use the Q-learning framework to search neural architectures, but the method is applied on cell-based search space. In this work, the NAS method is used to design a normal cell. The agent decides operations in the cell one by one. The state is defined as a tuple, including the number of operations in the current cell, and the information of the last operation. As stated before, unary and binary operations are equally treated in this search space. Thus the information of operations contains the type of operation, kernel size, and one or two input sources. To speed up the convergence of Q-values, rewards in the iterative process are given by reward shaping [91]. Therefore, the Q-values of iterative states are much higher in the early stage during training. To save time in the performance estimation process, the authors use an early stop strategy to get an approximate value. Besides, the authors study the correlation between performance and the architecture's flop and density, which can be used to give a more accurate prediction of the architectures.

Policy Gradient Methods
Policy gradient methods are alternative approaches in RL where the value functions do not appear and instead, the policy π_θ is directly learned by a collection of learnable parameters. These methods select actions without explicitly consulting a

value function. Instead, a controller is used to provide the distribution of actions given parameters θ in policy π. These parameters are continuous and usually updated by gradient descent methods. However, the rewards are not differentiable to these parameters as the performance depends both on the actions selected and the distribution of states after the action is taken. Hence, the empirical estimate of the gradient is needed here to perform updating. REINFORCE is a classical algorithm that estimates the gradient as:

$$\mathbb{E}_\pi \left[\sum_{t=0}^{T} \nabla_\theta \ln P(a_t|s_t, \theta)(r_t - b) \right], \tag{1.20}$$

where $P(a_t|s_t, \theta)$ is the probability of choosing action a_t at state s_t, and b is a predefined baseline reward to reduce variance. At the very beginning, the policy will give random choices at all states. After receiving a reward, the agent calculates the gradient and updates policy parameter θ to change the probability of choosing the current action in the same state. If the reward is bigger than the baseline reward b, the probability will increase, and vice versa. Therefore, after several episodes, the agent will have a larger probability to choose the action such that the corresponding reward is relatively big. At the end of the algorithm, the optimized policy is obtained. The agent can always choose the action with the maximum probability to get the maximal expectation of reward. Compared with Q-Learning, a main difference in policy gradient methods is that the balance of exploration and exploitation is controlled by the policy parameters but not an extra algorithm such as ϵ-greedy method. In policy gradient methods, the probability of choosing different actions is relatively similar. Via the learned policy parameters, actions with higher rewards are chosen more often, but actions with lower rewards also have a probability to be chosen. An advantage brought by policy gradient is that some prior knowledge can be modeled in the policy and influence the calculation of probabilities. Similar actions can be modeled by common parameters under the assumption that similar actions have similar rewards. Thus the backward process can affect a bunch of actions at the same states. Analogously, the relationship among different states can also be considered. This character show that the design of policy controller is of great importance.

Zoph and Le [123] apply policy gradient methods on NAS problem for the first time. The global search space with skip connections is adopted, and there are only convolution layers with different hyper-parameters in the space. The states and actions are defined in hyper-parameter level, i.e., several consecutive actions sequentially determine the hyper-parameters of an architecture's layer, including the height and width of filter and stride. An RNN controller is used to generate the action distributions, so later hyper-parameters can be determined based on previous hyper-parameters through the information passed by hidden states in RNN. As the last part of each layer, the RNN controller should decide which previous layers are the inputs of the current layer at a RNN cell, named anchor point, by the following probability:

$$P(j \rightarrow i) = \text{sigmoid}(v^T \tanh(W_{prev} \times h_j + W_{curr} \times h_i)),   \qquad (1.21)$$

where $j \rightarrow i$ means the j-th layer is an input of the i-th layer, and h_i represents the hidden state of controller at i-th layer. W_{prev} and W_{curr} are learnable parameters. All previous layers should be considered by the function at an anchor point. The inputs are concatenated before fed into the current layer. However, the generated architecture may not make sense since that some layers have no input or the image sizes of inputs of a layer do not match. To ensure the architectures are legal, the authors implement some regulations. First, if a layer has no input, the original image is used as its input. Second, the final layer aggregates all layer outputs that have no outcoming edges by concatenation. Third, if two inputs' image size do not match, the smaller one should be padded with zeros to achieve the same size with the larger one. These regulations are simple and practical. Other search spaces such as cell-based space have been applied by similar regulations as well. REINFORCE is used to update the RNN parameters. The probability of actions are implied by a softmax function.

Zoph et al. [124] adopt policy gradient methods in cell-based search space, where block structure with two operations is used to construct cells. Five consecutive actions of the agent determine all elements of a block, including two input sources, types of two operations, and combination method of the two results. The combination method candidates include addition and concatenation, but addition dominates the choices of combination methods in the experiments. Proximal Policy Optimization [98] is used to update the controller, since it is faster and more stable.

Cai et al. [58] redefine the decision making process of architecture construction to reuse weights of previous architectures, improving the efficiency in the performance estimation process. The states are still defined as the condition of the architecture. The actions contain widening one layer and inserting a new layer into the architecture to deepen the network. Both of them are performed by means of function-preserving transformations, i.e., the function being modeled remains unchanged although the structure of the architecture changes. Therefore, initializing new architectures by the weights of the previous architectures maintains the original accuracy. The performance can be obtained by fine-tuning new architectures with a few epochs. To control how to select actions, a controller with multiple parts is adopted. Firstly, a bidirectional RNN is used to encode the operations into embeddings. For widening action, hidden states are fed into a classifier network to decide whether each operation should be widened and what is the widened shape of it. For deepening action, the final hidden state is fed into an RNN to sequentially determine where should a new layer be added and the hyper-parameters of it. At each step, several architecture candidates are provided by the agent. Those architectures are evaluated by reusing previous weights, thus greatly save time during the performance estimation process. The controller is updated by REINFORCE. However, in this framework, the backbone of the architectures is determined manually, since there is no mechanism to explore the operation type choices and input source choices, limiting the ability to design more variable architectures.

Cai at el. [59] further improve the above work by using a path-level network transformation framework. Since a convolutional layer or an identity map can be converted to an equivalent multi-branch motif by splitting or replicating the input at the beginning and adding or concatenation at the end, the function-preserving action candidates contain transforming a node to a multi-branch motif, inserting an identity node, and replacing an identity node with a candidate operation. This definition of actions permits NAS methods to explore a tree-structured architecture space. The controller for action selection contains an LSTM and three classifiers. To handle this type of structure, tree-structured LSTM is used to encode the network. Then, for a node that has only one leaf node, a classifier determines whether it should be transformed to multi-branch motif. For a leaf node, a classifier determines whether a new identity node should be inserted as its son. For an identity node, a classifier determines whether it should be replaced by another operation. All parameters in the controller are updated by REINFORCE.

C. Evolution Algorithm

Evolutionary Neural Architecture Search (ENAS), generally speaking, is a subfield of NAS research area that uses Evolutionary Computing (EC) methods to optimize DNN architecture search instead of Reinforcement Learning and Gradient-Based methods. To draw a clear picture of this topic, we will first give a brief introduction of ENAS, and then introduce several representative works to exemplify the core ideas and challenges behind ENAS, followed by advanced topics including research on its efficiency and combinations of ENAS with other fields.

The idea of ENAS is first proposed in LargeEvo [96] by Google in 2017, showing competitive power of automatic neural architecture designs for image classification tasks. Since then, due to EC methods' nature of little human interference, insensitiveness to local minima, and no requirement for gradient information, ENAS has become a hot topic in NAS research field with interested researchers and works growing at a rapid speed [85]. Note that the idea of using EC methods to search neural network (NN) is not invented recently. It has been actually researched on decades ago in the name of neuro-evolution. ENAS distinguishes from neuro-evolution in that ENAS uses EC methods to search architectures and optimize the parameters of candidates using gradient-based methods, while neuro-evolution uses EC methods to tackle both problems. Another difference is that ENAS usually aims to search Deep NNs ,while neuro-evolution is for shallow ones. In this topic, we only focus on recent works on the concept of ENAS and omit ones in neuro-evolution.

To distinguish ENAS from other subfields of NAS, we will clarify several basic concepts and the evolutionary process commonly used in ENAS. An **individual** is a solution for ENAS, which is an encoded NN architecture candidate. Architectures are usually encoded into vectors by different **encoding strategies**, representing basic units contained, their corresponding parameters, and the connections between them. The basic units are all same as those introduced in previous sections, e.g., convolution layers, ResNet-Cells, etc. The **encoding space** of architectures defines all possible architectures to be found, which largely determines the algorithm's effectiveness and efficiency. Many constraints, e.g., fixed depth of architectures, partial fixed structure, etc., will narrow the encoding space, which may lead to architecture search with higher efficiency as well as lower effectiveness. At the beginning of the evolutionary process, a collection of individuals, a.k.a. **population**, is initialized and repeatedly goes through the **population updating** process consisted in evaluation, selection, and evolution. In practice, an individual is usually evaluated on validation dataset by task-specific metrics after being decoded back to an architecture, and the performance results, generally termed as **fitness**, are exploited to guide the selection of population. Note that fitness can also take into consideration costs of other objectives (e.g., the number of parameters) to comprehensively evaluate the architecture. Like natural selection, it is natural to endow individuals of better fitness with high probability to survive, and remove those of worse fitness from the population. The process of selecting survivals is also called environmental selection, followed by another selection, termed mate selection, in deciding which individuals to evolve. For simplicity, we call selection methods in both phases as **selection strategies**. Next, individuals selected will evolve through **evolutionary operators**, where the two most commonly used ones are mutation operator and crossover operator. The major difference between mutation operator and crossover operator is that the former usually modifies an individual based on its own while the latter exploits multiple individuals' information to make the change. A population after experiencing evaluation, selection, and evolution for one time is called a **generation**. After iterative updating for many generations until the stop criterion is reached, the last population is obtained to make up the final solutions.

Taxonomy of ENAS

To characterize ENAS algorithms, one can turn to the three core components of NAS, including search space (initial space, encoding space, and encoding strategies), search strategies (EC methods exploited, selection strategies, and evolutionary operations), and performance estimation strategy (evaluation objectives and evaluation efficiency).

By the EC methods being exploited, ENAS can be classified into three main aspects and further subdivided.

- Evolutionary Algorithms (EA)-based methods,which is the major part in ENAS. It can be subdivided into Genetic Algorithms (GA), Genetic Programming (GP), and Evolutionary Strategy (ES) methods.
- Swarm Intelligence (SI)-based methods are composed of Particle Swarm Optimization (PSO) and Ant Colony Optimization (ACO) methods.
- Others, including Memetic, Differential Evolution (DE), etc.

Among all the sub-aspects, GA is the dominant method utilized in ENAS.

From the perspective of encoding space, ENAS can be divided into 4 categories as NAS does:

- Layer-based encoding, where the basic units in the encoding space are primitive layers, such as pooling and convolution. Since a sophisticated architecture may contain a large number of primitive layers, this method is usually very time-consuming and inefficient.
- Block-based encoding, where various layers of different types are combined as blocks as basic units, such as ResBlock, DenseBlock, ConvBlock, and InceptionBlock. Some hyper-parameters of blocks, like the number of layers and sizes of kernels, are partially predefined so that the search space can be significantly reduced, which is a great step towards the trade-off between efficiency and effectiveness.
- Cell-based encoding, which can be viewed as a special case of block-based method. The algorithm first search several types of cells and assemble them by stacking or other connection graphs. It further reduces the search space, and shows promising results in recent works.
- Topology-based encoding, which only learns how to connect units without caring about how the unit is composed.

Initial space for ENAS refers to the first population composition in the evolutionary process, which can be trivial space, random space, or rich space. A trival space takes as simple architectures as possible for the first population and leave it all for the algorithm to gradually search more complex and better architectures from scratch, which may make it hard to search the optimal but in potential novel ones. A rich space, on the contrary, lets the algorithm start from existing sophisticated architectures so that it can learn to improve or prune already well-performed architectures while may lead to less novelty.

In terms of encoding strategies, ENAS can take in fixed-length encoding or variable-length encoding methods, and linear or nonlinear connections within architectures. The variable and nonlinear ones make the algorithm more flexible and powerful, while it also becomes more complex and expertise demanding.

Selection strategies of five main kinds can be used in ENAS:

- Elitism, which keeps alive individuals with better fitness.
- Discarding the worst, which kills individuals with worse fitness.
- Roulette wheel selection, which endows each individual with a surviving probability in proportion to its fitness.
- Tournament selection, which samples equally from individuals and selects ones of best fitness for several times.
- Others. Some may consider not only a individual's own fitness, but also its contribution to the diversity of population selected so that the algorithm can be insensitive to local optima. For instance, the authors of [71] propose to prefer individuals of lower density and higher distance.

A variety of recent works focus on evolutionary operations, evaluation objectives, and evaluation efficiency, since they are either expertise-demanding for adopting EC to search DNN architectures, or of great importance to effectiveness, efficiency and practicality in real-world applications.

For evolutionary operations, some works research on how to allow the encoded information to mutate within a given range, including polynomial mutation [101], Gaussian mutation [86], RNN mutation [90], etc. While the former two give a parameterized distribution like polynomial or Gaussian distribution upon mutation range, RNN mutation exploits model parameters and its performance, and learns to guide the mutation range step by step. Distinct from traditional uniform distribution, these distributions may keep ENAS off wasting time on worse mutations. To further save computing resources for evolution, researchers propose function-preserving operators, namely network morphism, which changes the architecture without loss of acquired experience by forcing mutated individuals to be impossible to act worse than their parents [106]. Other than efficiency challenges, some works focus on maintaining post-mutation diversity to avoid local optima and guarantee the effectiveness of search [78, 94, 103].

By evaluation objectives, ENAS can be divided into single-objective and multi-objective. Single-objective ENAS usually takes as objective only one task-specific metric, e.g., classification accuracy, while multi-objective ENAS may aim at searching architectures optimal for multiple objectives, including performance metrics, hardware computational costs, model size, etc [71, 87, 88, 114]. Architectures searched with multi-objective can be more adaptive for target application environments, e.g., a much lighter model with little loss of performance may be suitable for applications in mobile devices.

As for evaluation efficiency, a variety of methods are dedicated to reducing computation costs in evolutionary process and architecture evaluation.

- Weight inheritance. Some parameters of the evolved individuals can be inherited from their parents for those unchanged partial structures [96]. Since some evolutionary operators may not disrupt the whole architecture, this method can save amounts of time compared with training from scratch.
- Reduction in iteration-related parameters, including early stopping policy, population size reduction, and training dataset reduction. To reduce time cost, they respectively kill the weak individuals early to reduce training epochs, reduce population size, or use a smaller but similar dataset to act as evaluation proxy. However, they may lead to inaccurate estimation about individuals' fitness or trap in local optima, causing worse architecture search performance.
- Efficient systems. Some propose distributed asynchronous system [75, 105] and hardware-aware algorithms [65, 66] for efficient computing.

To illustrate core parts of ENAS concretely, we next take LargeEvo [96] as an example and briefly introduce some other representative works in ENAS.

Representatives in ENAS
LargeEvo [96] is viewed as the first work in ENAS, which aims at minimizing human participation in architecture designs for image classification tasks. Specif-

ically, it adopts GA method, trivial initial space, layer-based encoding space, tournament selection strategy, and weight inheritance for evaluation efficiency.

An individual is encoded as a directed acyclic graph (DAG), where a vertex represents activation of image data including ReLU and plain linear units, and an edge represents identity connections or convolution layers. Image data, represented by rank-3 tensors (X-axis, Y-axis, Channels), is activated by the activation layer assigned in the vertices, and then be used as inputs for connection layers of outedges. Data from edges incident on one vertex is aggregated to continue flowing through calculations of the vertices and edges till reaching the vertices with zero outdegree so that the final outputs are obtained. Each initial individual is composed of just a single-layer model without convolutions, and the population gradually grows to complexities during the evolutionary process.

In the selection phase, random pairwise individuals are chosen for competitions repeatedly until the population size is reached. The winners of the tournament will experience the mutations to evolve to the next generation. Mutation operators include modifying parameters (learning rate, stride sizes, channel sizes, filter sizes, weights) and modifying connections (add or remove skip connection or convolution layer). Note that weights of architectures are optimized by gradient-based methods, and mutation of weights is to reset architecture weights inherited from parents so that the model can be trained again to escape local optima. Modification on parameters will result in a new value from uniform distributions around the original value, and for connections, pairwise vertices are also randomly chosen. The search space is unbounded since connection insertions may cause the model to grow in size.

Each architecture is trained to completion in the evaluation stage to get an accurate estimation of individuals, while it makes the evolutionary process unaffordably time-consuming. To preclude individuals from training from scratch, it exploits weight inheritance whenever possible, e.g., weights of one convolution layer are inherited if its related parameters are completely unchanged, such as sizes of stride, filter, and channels.

Though simple, LargeEvo takes the first step in the development of ENAS, showing the potential of combining EC methods with NAS. While showing competitiveness with architectures searched for image classification tasks, it is highly demanding for computational resources, making it unpractical in real-world applications. Since then, many works aims to reduce the computational burden by exploiting more compact encoding space, weight sharing, etc. We will next introduce the LEMONADE algorithm as one representative of those works, which significantly reduces GPU days for training on CIFAR-10 from LargeEvo's 2750 days to only 80 days while achieving better performance.

LEMONADE [71] is a GA-based method aimed at reducing computational costs in automatically architecture search in the area of image classification. Specifically, to improve the efficiency of searching, it adopts block-based encoding space, network morphism for weight inheritance, and multi-objective optimization to consider computational costs. In the following, we brief methods associated with efficiency improvement while omitting others.

The basic unit of the search space is only one block, composed of convolution layer, batch normalization, and ReLU activation sequentially. During the evolutionary process, it searches architectures by changing the number of blocks, connections between blocks, and the number of filters within blocks. The block design is de facto in sophisticated hand-crafted architectures in image tasks. Therefore, the search space is greatly narrowed and time wasted in searching for largely unpromising combinations of basic layers is saved.

Network morphism is utilized as mutation operators so that during evaluation individuals the models can inherit all weights from parents and avoid training from scratch. Network morphism is a network operator that does not change all the outputs from architectures after they are modified by the operator. For example, a fully connected layer, whose weights are initialized as an identity matrix, can be added after a model to make the model deeper while keeping outputs unchanged. In this case, the modified model is initialized with all weights from parents plus an identity matrix to be trained, and it takes much less time for the model to obtain optimal weights than models with random weight initialization. Several network morphism strategies are proposed to make the architecture deeper (add blocks), wider (add the number of filters within blocks), and more flexible (add skip connections). However, keeping model growing in sizes, as done by network morphisms, can increasingly improve the performance of architectures while going counter to the purpose of reducing computational costs. To this end, several approximate network morphism algorithms are proposed to decrease model sizes, including making model shallower (remove a convolution or skip-connection), narrower (remove parts of filters within blocks), etc. These operators make effects of modifications upon changes in model outputs as slight as possible by retraining several epochs only for layers affected, e.g., the removed layer and its direct predecessor and successor. This enables modified models to quickly reach optimal weights as well.

A multi-objective evaluation paradigm is utilized so that other objectives like computational costs can be taken into consideration and the need for evaluation of different objectives can be reassigned according to their computational costs to save time. By time costs, objectives are classified into cheap ones (like the number of parameters of models) and expensive ones (like performances on datasets). The cheap ones are utilized to provide distributions for the population so that individuals can be selected with methods controlling post-mutation diversity, like choosing individuals with probability in proportion to reciprocal of density. The expensive ones are evaluated much fewer times than cheap ones. After evaluation, non-dominant individuals are retained to the next population, where non-dominant defines an individual that is not all worse than other individuals. In this way, the obtained generation is an approximation of the Pareto front of multi-objectives. Since it is an anytime algorithm, one can choose a variety of models based on one's trade-off between several objectives such as performance and computational costs.

While LEMONADE greatly reduces searching time for CIFAR-10, 80 GPU days in need for such a toy-level dataset are still prohibitively expensive for end-users.

We will next brief another work that continues improving the efficiency of ENAS in image tasks.

Similar to LEMONADE, NSGANet [88] is also a multi-objective method and search for the Pareto front of architectures by several objectives including performance metrics and computational costs. It distinguishes itself from others by utilizing different evolutionary operators and Bayesian learning to make search more efficient, which enables searching better architectures within 8 GPU days for layer-based space and 4 GPU days for cell-based space. We will next briefly introduce the basic ideas of their contributions to efficiency.

An individual is a sequence of blocks, each of which encodes a directed acyclic graph by binary bits. The search process includes the two stages of exploration and exploitation, where the former takes the traditional evolutionary operators including mutation and crossover, while the latter utilizes the Bayesian network to make use of search history by calculating conditional distributions for individual evolution. The crossover operator keeps common bits shared between two selected parents and recombines distinct ones to generate new individuals. Weight inheritance is used for common bits and model complexity is maintained by restricting the number of '1' bits to lie between ones in both parents. The mutation operator flips only bits of the individual so that weight inheritance can be also exploited. Different from the exploration stage to get diverse populations, the exploitation stage is aimed at exploiting searching experience and reinforce patterns commonly shared among successful architectures searched in history. It calculates conditional distributions of pairwise blocks and takes them as Markov transition distributions to generate new individuals, where individual generation is viewed as a sequential decision problem. For instance, the probability of architecture (x_0, x_1) to be generated, denoted as $p(x_0, x_1)$, is $p(x_0) \times p(x_1|x_0)$, where x_i refers to a block encoding in the architecture. All conditional distributions x_i are estimated by using the population history and updated throughout the exploitation stage. Evidenced by ablation studies shown in the paper, the Bayesian part outperforms random search, which exploits the history of exploitation and greatly saves time.

Combinations with Other Fields Differentiable NAS dominates the NAS research area recently for its high searching efficiency, which usually first trains a SuperNet and optimizes architectures using gradient descent methods. However, it is shown that searched architectures suffer from a lack of variety. To this end, Continuous Evolution for Efficient Neural Architecture Search (CARS [114]) proposes to combine Genetic algorithms with differential NAS to search architectures with both diversity and efficiency. Similar to DARTS, a SuperNet is proposed to contain all possible architecture parameters and connections within search space, and individuals are encoded by weights and connection masks upon SuperNet. For the convenience of optimization in continuance, the mask is defined as a sequence of continuous values lying within zero to one. An individual can be decoded into architecture by taking each mask to weight corresponding connections to get continuous one during training or choosing connections of max values in masks to get discrete one during testing. Two stages are alternatively adopted in the search process, i.e., parameter optimization and architecture optimization. During parameter optimization, a mini-batch of individuals in the current population is used to estimate parameter gradients for optimizing architecture weights, and since all weights are shared in the SuperNet, weight inheritance is naturally carried out so that individuals can obtain the optimal parameters much faster than training from scratch. During architecture optimization, as done by common multi-objective NAS such as LEMONADE, CARS adopts the NSGA method, which selects non-dominant individuals to obtain next generation so that each generation falls onto Pareto front searched so far. However, the traditional NSGA method may not be suitable for the selection of population in SuperNet because of a small model trap phenomenon. That is, in the selection stage, parameters of SuperNet may still need optimization and thus the accuracy of each individual may be not well estimated, especially for those large models with slower convergence and better performance, which causes that the algorithm usually prefers small models and kills large models in the early stage. To this end, CARS proposes a method called pNSGA-III for protecting those well-performed large models. It obtains the first Pareto front only based on accuracy and number of parameters while then the second Pareto front only based on the number of parameters and the increasing speed of accuracy. Two populations are then merged and top-K individuals in the merged union set are kept to the next generation. In this way, well-performed large models can be protected thanks to the metric of increasing speed of accuracy. The experiments show that compared to LEMONADE, CARS can find models with better performance and fewer parameters using much less time (only 0.4 GPU days) in CIFAR-10.

Before the advent of differentiable NAS, EA and RL methods are dominant in NAS and both have achieved state-of-the-art results in a variety of tasks. However, both of them have limitations respectively. Generally, solutions by EA-based NAS are guaranteed to be diverse while the population evolution process can be inefficient, since evolutionary operators that randomly modify the population are predefined and unchanged throughout the process. For RL-based NAS, architectures are usually generated by determining units one by one and trained till optimized so that decisions made by RL are to be more sophisticated with more trial and error. However, a large number of trials are needed as supervision for training decision making of each state, which also causes efficiency problems for RL-based methods. To tackle problems and exploit the advantages of both methods, Reinforced Evolutionary Neural Architecture Search (RENAS [64]) proposes to combine two methods by replacing predefined evolutionary operators with an RL controller. Specifically, the controller consists of an encoder, mutation-router, and mutators for operations and connections, which learns to select what parts of one individual to mutate and how to mutate them. Adopting cell-based space, the method defines each cell as a DAG where each node takes in two input feature maps (i_1, i_2) and two corresponding operations (o_1, o_2). The encoder is a bidirectional recurrent network that encodes input connections and operations into hidden states for each unit in cells. The mutation-router, composed of a fully connected layer and softmax function, chooses which one of (i_1, i_2, o_1, o_2) to mutate by sampling based on probabilities given by outputs after softmax. If the mutation is to fall upon (i_1, i_2), another neural network learns to select which nodes before the current nodes to connect. If the mutation is for operation, then the mutated operation will be sampled from operation sets based on the hidden state of the node. During the evolutionary process, the controller's parameters will keep being updated via policy gradient. In this way, the mutation in the whole process will not stay unchanged, but instead, it adjusts itself to the trade-off between exploration and exploitation. Meanwhile, the RL controller does not need to decide the whole architecture step by step, while only need to decide how to mutate and improve the current individuals, which improves training efficiency. The results show that in performance and efficiency, RENAS exceeds NASNet and AmoebaNet, which are representatives in RL-based NAS and EA-based NAS respectively.

D. Surrogate Model-Based Optimization

Surrogate Model-Based Optimization (SMBO) methods use a surrogate model $\hat{f}$ to approximate the response function f in Eq. (1.14). The surrogate $\hat{f}$ is usually a machine learning model learned by minimization of value loss with f. Formally,

$$\min \mathcal{L}(\hat{f}(\alpha), f(\alpha)). \tag{1.22}$$

The architectures used to learn the surrogate are sampled from a certain distribution. $\mathcal{L}$ can be directly calculated by square loss to learn an accurate surrogate function for performance estimation, or a loss related to performance ranking is used since NAS only cares about the architecture with the best performance. With a surrogate function, we can efficiently give architecture candidates by $\arg\max_\alpha \hat{f}(\alpha)$. After these candidates are evaluated, the corresponding performances are used to update the loss in Eq. (1.22) and optimize $\hat{f}$. These steps are executed until a convergence criterion is reached.

In this type of NAS method, the modeling of architecture space is contained by the design of the surrogate model, and the updating process is gradient descent methods to minimize the loss. Some of the SMBO-based methods use RNN as the surrogate to model the architecture space, similar to some RL-based methods. The main difference between them is the RNNs in RL-based methods are used to model the action distribution of each step, while those in SMBO-based methods are used to predict the performances. Besides, different from RL where an architecture distribution is modeled, or EA where a new population is generated by evaluation actions, SMBO-based methods need an extra mechanism to provide architecture suggestions. The simplest way to provide the suggestions is the full search or random search by the surrogate model. Evolution actions such as mutation can be also used to explore the neighborhood of existing architectures. Thus, the border between SMBO-based methods and other types of NAS methods is somehow vague. Methods exploring the search space by EA and evaluating architectures with a surrogate model can be either regarded as an SMBO-base model or EA-based methods with the surrogate as a performance estimation strategy. Because of the vague characterization, the methods in this section may also belong to another class of NAS methods. Besides, other search strategies combined with a surrogate model as a performance estimation strategy also belong to SMBO-based methods to some extent.

Liu et al. [81] incorporate a surrogate model and search for architectures in the cell-based space. In this paper, an LSTM is used as a surrogate to predict the performance of architectures since it can handle variable-sized inputs. $4b$ elements are fed into the LSTM, where b is the number of blocks in a cell. Each block has four one-hot vectors indicating the two input sources and the types of two operations. The final hidden state of the LSTM does through a fully connected layer and sigmoid function to regress the valid accuracy. The generation of architectures is like a beam search process. They explore the search space by progressively increasing the number of blocks in a cell. They begin the search process by generating architectures with only one block in each cell. At each step, a new set of candidates is generated by expanding the current pool of cells with an additional block and the top k of those given by the surrogate model are trained. Then, the surrogate model is updated by considering the newly obtained k architectures and their corresponding performances. In detail, the surrogate is trained by the new data using a few steps of SGD. Besides, since the sample size is very small, they fit an ensemble of 5 predictors, each fit 4/5 of all the data available at each step of the search process. This mechanism empirically reduces the variance of the predictions.

Kandasamy et al. [76] tackle the problem with Bayesian optimization. Bayesian optimization methods use a combination of a probabilistic surrogate model and an acquisition function to obtain suitable candidates. The acquisition function measures the utility by accounting for both the predicted response and the uncertainty. In their approach, the surrogate is modeled by a Gaussian process and expected improvement is used as an acquisition function. To handle the architecture space, the authors redefine a kernel function to estimate the distance between two networks via an optimal transport program considering the network topology, the location, and frequency of operations as well as the number of feature maps. The redefined distance is the minimum of a matching scheme that attempts to match the computation at the layers of one network to the layers of the other. Penalties for matching layers with different types of operations or those at structurally different positions are incurred. Based on the distance definition, the acquisition function can be calculated. The authors use the EA approach to explore the search space and optimize the acquisition function. The approach begins with an initial pool of networks. Then a set of mutation candidates are sampled from the pool with the probabilities corresponding to their acquisition, i.e., architectures with higher acquisition have more chances to be selected. Next, those candidates are modified by some transformations to generate new architectures. Finally, those new architectures are evaluated and put into the pool, and the surrogate is updated. All these steps are repeated to get the optimal architecture.

Shi et al. [99] combine Bayesian optimization with GCN as a surrogate model. In BONAS, the neural architecture is seen as a graph, each operation is given a one-hot vector as its node feature. A global node that is connected with all other nodes is added to the graph to aggregate features of the whole graph. They apply GCN on the architecture graph and use the graph embedding to predict the performance of the architecture by a single-hidden-layer network with a sigmoid function. The exploration process is similar to NASBOT, where an evolutionary algorithm is used.

Luo et al. [89] propose an SMBO-based method NAO which explores in the embedding space. To transform architecture into a vector, they use an autoencoder to encode architectures. Besides, an MLP is used as a surrogate model which uses continuous embeddings to predict performances of architectures. The autoencoder and the surrogate model are simultaneously trained, i.e., the reconstruction loss of the autoencoder and the square loss of predicted and ground-truth performances are combined. In detail, they use the NASnet space, which is a cell-based space using a block structure. The encoder and the decoder are both sequence-to-sequence models using LSTMs. In the encoder, the information of blocks is fed into the encoder one by one and a series of hidden state vectors are generated. The decoder uses the vector series to reconstruct each of the blocks. The surrogate adopts average pooling to the vector series and puts them into MLP. A novel point in this work is that they use a gradient-based method to explore the architecture space. The architecture pool is initialized by a set of architecture candidates, and the autoencoder and surrogate are trained to obtain their corresponding performances. At each turn, the architectures with top k performances in the pool are chosen. Their embeddings are changed by moving one step forward, whose distance is predefined, in the direction of the gradient to maximize the predicted performance given by the surrogate. Then the new embeddings are fed into the decoder to generate new architectures, which will be evaluated and put into the pool. In case that the newly decoded architecture is the same as some previous architecture, the corresponding embedding will keep changing through moving towards more steps until decoding a different architecture. Finally, the surrogate model is updated by the new architectures. The above steps are repeated.

Zhang et al. [118] also use an autoencoder framework. Differently, they regard architectures as graphs and use a variational graph autoencoder to transform the architectures into a continuous latent space. The graph autoencoder is trained as a pretreatment process. Another novel point in this paper finds the problem that since only one architecture is decoded at one turn by a step of gradient descent on the latent space, it is easy to get a locally optimal point according to the rich-get-richer problem on the super network. Thus it encourages the exploration of a larger architecture space by a probabilistic exploration enhancement method. In the search process, the model maintains an architecture pool A, where are filled with visited architectures in previous steps. To encourage exploration on new latent space, the hidden embedding is updated by:

$$\alpha_\theta^{(i+1)} = \alpha_\theta^i - (1 - \gamma)\nabla\mathcal{L}_{val}(\alpha_\theta^i, W) - \gamma\nabla N(\alpha_\theta^i, A). \tag{1.23}$$

Here, $N(\alpha_\theta^i, A)$ is to measure the novelty of architecture α_i from the pool, λ is a hyper-parameter to control the balance between exploration and exploitation. At each turn, the model decodes α_θ to get an architecture, which is used to update the weights on the super network. Then the hidden embedding is updated by the above equation. At the end of the search phase, the final hidden embedding is decoded to get the optimal architecture.

E. One-Shot Architecture Search

One-shot architecture search methods train a single super network that contains all possible architectures in the defined search space. All training and evaluation processes of architectures are performed on the same super network. The parameters of the same operation at the same position are shared by all architectures with the same layer. This idea comes from ENAS [93], which is an RL-based NAS method using the super network to evaluate the performances of architecture candidates. In an iteration of RL, there is a super network training procedure, when a set of architectures used for training are sampled from the RL controller. Then a forward propagation process is adopted on the super network where only the parameters of layers existing in these architectures are used. In the following backpropagation process, only the used parameters are updated by the optimizer, other parameters are kept unchanged. When architecture candidates need to be evaluated in the performance estimation procedure, they are adopted on the super network by a forward propagation to get the approximate performances. Thus, architecture search methods with this performance estimation strategy have relatively low search effort which is only slightly greater than the training costs of one architecture in the search space. After ENAS, many works propose specialized search strategies based on the super network. This type of NAS method has only one performance estimation strategy which is using the shared parameter on the super network to evaluate. This family of methods design the architecture space exploration mechanism highly connecting to the super network.

There are mainly two types of methods modeling on the super network. One is using a mixed operation. All possible operations are mixed at each layer by a weighted sum. The forward propagation process on the super network is performed on a mixture of multiple architectures. The other is using operation distributions on all layers. This model is similar to RL-based models where an architecture distribution is modeled in the RL controller. Only one path of a single architecture is used in each forward propagation process. The main difference between operation distribution on the super network and RL-based models is that the operation distribution on the super network is a joint distribution, where operations are not sequentially decided.

DARTS [84] is a representative NAS method with both conciseness and effectiveness, serving as the exploration object in this chapter. In DARTS, neural architecture is represented as a directed acyclic graph (DAG) where the input data is the source node, the output is the sink node, and the edges represent operations (e.g., a GCN layer, or a linear layer) adopted on the data. Following a micro search space setting,

Fig. 1.8 An example of
super network

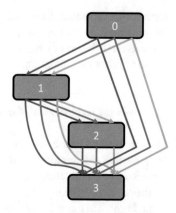

DARTS only searches which operation should be selected and how the nodes in the DAG should be connected. The calculation inside each operation is predefined.

There are two phases in DARTS procedure, the searching phase and the evaluation phase. In the searching phase, a super network (shown in Fig. 1.8) is constructed, where edges exist between each two nodes, i.e., $e_{i,j}$ exists, $\forall 0 \le i < j \le N - 1$. Each edge is a mixed operation that can be calculated by

$$e_{i,j}(\mathbf{x}_i) = \sum_{o \in O} \frac{exp\{\alpha_{i,j}^o\}}{\sum_{o' \in \mathbf{O}} exp\{\alpha_{i,j}^{o'}\}} \cdot o(\mathbf{x}_i), \qquad (1.24)$$

where $\mathbf{x_i}$ is the output of node i, O is candidate operation set, and α is learnable architecture parameters. After calculating the mixed weighted sum, each node aggregates all input edges by

$$\mathbf{x}_j = \sum_{i < j} e_{i,j}(\mathbf{x}_i). \qquad (1.25)$$

As such, all possible operations at all possible positions are contained in the super network. Besides, fewer cells are used to construct the super network in searching phase. DARTS uses gradient based methods to optimize both architecture parameters α and operation parameters by a bi-level optimization scheme. Since the parameters are all trained in the super network, W of different architectures are shared with each other, the formulation of Eq. (1.14) is changed to

$$\min_{\mathcal{A}} \quad \mathcal{L}_{val}(W^*, \mathcal{A})$$
$$s.t. \quad W^* = \mathrm{argmin}_W \, \mathbb{E}_{\mathcal{A} \in \Gamma(\mathcal{A})} \mathcal{L}_{train}(W, \mathcal{A}). \qquad (1.26)$$

Here, $\Gamma(\mathcal{A})$ is architecture distribution learned in the searching phase. In this formulation, W and $\mathcal{A}$ are independently learned, improving the search efficiency.

At the end of the training procedure, the two operations with the largest two α of each node are chosen, and the selected operations compose the optimal architecture. When it comes to the evaluation phase, a new network based on the designed architecture is constructed. This network will be trained from scratch and finally get tested after retraining. In DARTS, a significant difference is that there are no architecture candidates generated and no explicit performance estimation of architecture candidates in the search phase. Instead, a mixed architecture is trained which can be regarded as to approximate the architecture candidate. The loss of mixed architecture can be seen as the feedback, and the updating process is backward propagation to the mixture weights on edges. Since the weights of mixed operations are differentiable, gradient descent methods can be directly adopted to optimize those weights. However, there are also some problems in DARTS.

- All parameters of all operation candidates must be kept in memory all the time. Since the super network covers the entire search space, there is extremely heavy memory usage.
- DARTS does not directly optimize the objective function in Eq. (1.14). Instead, weights of mixed operations are optimized during the searching phase. The architecture discretized by those weights has a gap with the mixture architecture, which may cause a performance drop in the evaluation phase.

Using operation distribution instead of mixed operation is one of the solutions to the first problem. SNAS [110] uses concrete distribution to relax the operation distribution to be continuous and differentiable with the reparameterization trick. The mixture operation in DARTS can be reformulated as:

$$e_{i,j}(\mathbf{x}_i) = \mathbf{z}_{i,j}^T \cdot o(\mathbf{x}_i), \tag{1.27}$$

where $\mathbf{z}_{i,j}^T$ is a vector indicating the weights of operation candidates. To save computational resources and keep the objective function the same as that of ENAS and other RL-based methods, they sample the operations on each edge instead of using a mixture operation as in DARTS. Thus $\mathbf{z}_{i,j}^T$ becomes a one-hot vector. The gradients are estimated based on Monte Carlo samples. However, the one-hot vectors are not differentiable to the architecture parameters. To get the gradient of architecture parameters, concrete distribution is used here. Let

$$\mathbf{z}_{i,j}^k = f_{\alpha_{i,j}}(G_{i,j}^k) = \frac{exp((\log \alpha_{i,j}^k + G_{i,j}^k)/\lambda)}{\sum_l exp((\log \alpha_{i,j}^l + G_{i,j}^l)/\lambda)}, \tag{1.28}$$

where $z_{i,j}$ is the softened one-hot random variable for operation selection at edge (i, j). $G_{i,j}^k = -log(-log(U_{i,j}^k))$ is the k-th Gumbel random variable, $U_{i,j}^k$ is a uniform random variable, $\alpha_{i,j}$ is the architecture parameter, λ is the temperature of the softmax which is steadily annealed to be close to zero in the searching process.

Then the gradient to architecture parameters can be estimated as

$$\frac{\partial \mathcal{L}}{\partial \alpha_{i,j}^k} = \frac{\partial \mathcal{L}}{\partial x_j} o_{i,j}^T(x_i)(\delta(k'-k) - z_{i,j}) z_{i,j}^k \frac{1}{\lambda \alpha_{i,j}^k}, \tag{1.29}$$

which is named search gradient. Optimizing with this gradient is equivalent to the policy gradient methods in RL. Compared with RL-based model where the architecture construction is modeled as a sequential decision making process, joint distribution of the whole architecture is modeled in the super network in this work. Hence, it is more efficient to search for the optimal architecture. Since architectures are sampled from the distribution during the training procedure of the super network and architecture candidates evaluation, the memory only holds the backpropagation process of the weights in the selected layers of those architectures, saving lots of computational resources. GDAS [70] and ProxylessNAS [60], two works in the same period, use similar modeling methods to reduce memory overhead of super network.

In DSNAS [73], the memory overhead is further reduced. Although backpropagation of operation parameters that are not selected is not implemented, those parameters are still kept in memory to calculate the search gradient. DSNAS proposes a new gradient calculation method that does not need to keep those parameters. By multiplying a dummy variable after each feature map, the gradient to architecture parameters can be estimated by $\frac{\partial \log p_\alpha(\mathbf{Z})}{\partial \alpha} \frac{\partial L}{\partial 1^{dummy}}$. Since the memory overhead is largely reduced, works like ProxylessNAS and DSNAS do not use a proxy model in the search phase. Instead, they directly search on the super network with the same size as the target architecture.

PCDARTS [111] reduces the computational overhead in another way. The mixed operation calculation in PCDARTS is changed to:

$$e_{i,j}(\mathbf{x}_i) = \sum_{o \in O} \frac{exp\{\alpha_{i,j}^o\}}{\sum_{o' \in \mathbf{O}} exp\{\alpha_{i,j}^{o'}\}} \cdot o(\mathbf{S}_{i,j} * \mathbf{x}_i) + (1 - \mathbf{S}_{i,j}) * \mathbf{x}_i. \tag{1.30}$$

Here, $\mathbf{S}_{i,j}$ is the mask vector on the edge (i, j). The dimension of selected channels is assigned by 1, and the masked ones are assigned by 0. In practice, the proportion of selected channels is set to $1/K$, thus the memory overhead is reduced by K times. Besides, since each node x_j needs to pick up two input nodes from its precedents at the evaluation phase, but the architecture parameters are optimized by randomly sampled channels across iterations, introducing instability to the optimization process. This could cause undesired fluctuation in the resultant network architecture. To mitigate this problem, they introduce edge normalization by replacing Eq. (1.25):

$$\mathbf{x}_j = \sum_{i<j} \frac{exp(\beta_{i,j})}{\sum_{i'<j} exp(\beta_{i',j})} \cdot e_{i,j}(\mathbf{x}_i). \tag{1.31}$$

Here, $\beta_{i,j}$ is the parameter indicating the weight of edge (i, j). The connectivity is determined by both α and β. Since β is shared through the training process, the learned network architecture is insensitive to the sampled channels across iterations, making the architecture search more stable.

PDARTS [63] improves DARTS from another perspective. The gap between the super network in the searching phase and the discretized architecture in the evaluation phase are harmful to the performance of DARTS. A significant reason is that the super network is a proxy model, which is shallower than the discretized architecture. In DARTS, the super network has 8 cells, but discretized architecture on CIFAR-10 has 20 cells. This problem is named as depth gap. To verify it, they execute the search process of DARTS multiple times and find that the normal cells of discovered architectures tend to keep shallow connections. Thus they take the strategy that progressively increases the network depth during the search process so that at the end of the search, the depth is sufficiently close to the setting used in the evaluation. There are also difficulties in adopting the above methods. One is that the computational overhead increases with the depth of the network. To address this problem, they propose a search space approximation scheme to progressively reduce the number of candidate operations at the end of each stage, referring to the weights of operations in the previous state as the criterion of selection. The other problem is that when searching over a deeper architecture, the differentiable approaches tend to bias towards the skip-connect operation because it leads to the fastest gradient descent, limiting the representation ability of searched architectures. To this end, they propose search space regularization, which adds operation-level dropout to restrict the number of preserved skip-connects for further stability.

Many works take efforts on the second problem of DARTS, the gap between architectures in the searching and evaluation phase. DATA [61] changes the discretization process in DARTS by expanding the architecture space. In the discretization process of DARTS, a node only keeps the two operations with the largest two architecture parameters. However, in DATA, a node can have an arbitrary number of input edges. Moreover, multiple operations can be chosen at each edge. As such, the architecture space is expanded. To handle this search space, DATA uses an ensemble Gumbel-softmax estimator. As shown in Eq. (1.27), $\mathbf{z}_{i,j}^T$ is a one-hot vector in SNAS. Here, $\mathbf{z}_{i,j}^T$ is permitted to be any binary vector, noted as $\mathbf{b}_{i,j}$, i.e., $\mathbf{b}_{i,j} \in \{0, 1\}^K$. To sample the binary vectors, the estimator ensembles the results of multiple Gumbel-softmax estimator. Let $\mathbf{z}^{(1)}, \mathbf{z}^{(2)}, \cdots \mathbf{z}^{(M)}$ be the one-hot vectors sample by M Gumbel-softmax estimator, the binary code $\mathbf{b} = [b_1, b_2, \cdots, b_K]$ is ensembled by

$$b_k = \max_{1 \leq i \leq M} z_k^{(i)}, 1 \leq k \leq K. \tag{1.32}$$

That is, for each edge, the multiple Gumbel-softmax estimators choose one operation candidate for K times, then an operation is chosen in the ensemble Gumbel-softmax estimator if it has been chosen at least one time.

SGAS [79] improves DARTS by sequentially discretizing the super network in the searching phase. This behavior is somehow like the search space approximation scheme in PDARTS, where the operation candidates are reduced in different stages. SGAS apportions the architecture discretization process in many decision processes in the searching phase. In the decision process, a chosen edge is discretized to the operation with maximum weight. At the end of the searching phase, all edges have been discretized, so there is no gap between the searching and evaluation phase. SGAS proposes a criterion to determine which edge is chosen in a decision epoch by considering edge importance, selection certainty, and selection stability.

MiLeNAS [72] demonstrates that gradient errors caused by approximations used in bilevel optimization lead to suboptimality. Therefore, the authors reformulate the NAS problem as a mixed-level problem:

$$\min_{\alpha, w}[\mathcal{L}_{tr}(w * (\alpha), \alpha) + \lambda \mathcal{L}_{val}(w * (\alpha), \alpha)]. \tag{1.33}$$

Here, λ is a non-negative regularization parameter that balances the importance of training loss and validation loss. Compared with regular bi-level formulation, this kind of formulation can alleviate the overfitting issue and search for architectures with higher accuracy. The authors apply the first-order method to solve the mix-level optimization problem by:

$$w = w - \eta_w \nabla_w \mathcal{L}_{tr}(w, \alpha), \tag{1.34}$$

$$\alpha = \alpha - \eta_\alpha (\nabla_\alpha \mathcal{L}_{tr}(w, \alpha) + \lambda \nabla_\alpha \mathcal{L}_{val}(w, \alpha)). \tag{1.35}$$

In ISTA-NAS [113], the authors argue that the architecture space should be sparse. However, the joint distribution on the super network represents a dense space. To this end, the authors formulate NAS as a sparse coding problem. They perform the differentiable search on a compressed lower-dimensional space that has the same validation loss as the original sparse solution space and recover an architecture by solving the space coding problem. Assume the sparse coding space is B, and $\mathbf{b_j}$ is a vector in B representing the architecture weight of node j. The architecture weights can be derived by:

$$\mathbf{z_j} = \arg\min_{\mathbf{z}} \frac{1}{2} \| A_j \mathbf{z} - \mathbf{b}_j \|_2^2 + \lambda \| \mathbf{z} \|_1, 1 < j \leq n. \tag{1.36}$$

The super network is constructed by all these architectural weights. The optimization problem to W and $\mathbf{z}$ is converted to optimization of W and $\mathbf{b}$.

1.2.3   Performance Estimation Strategy

Performance evaluation of architectures is a core part of the process of neural architecture search, which usually costs a majority of the time, since architecture candidates, especially for deep and complex models, may need numbers of training epochs to obtain a well-estimated performance. For those early works of NAS, prohibitively high demands of computational resources make them impractical for users in industrial applications. To this end, performance estimation strategies have become a hot research topic in the field of NAS. How to speed up the performance estimation process while retaining its efficacy? Precious researchers give answers from four ways as follows:

- Lower fidelity estimates, which try to reduce the parameters related to evaluation time, including training epochs, dataset size, etc. This kind of estimation method is based on the assumption that the ranks of candidates are consistent before and after reducing fidelity so that the effectiveness of any search strategy only relying on ranks instead of absolute performance will not be disrupted. However, recent results indicate that this assumption may be broken when the gap between low-fidelity estimates and high ones is too large [117].
- Performance predictor, which usually learns a performance estimator based on the history of a full evaluation of previous architectures, to obtain performance estimates of candidates instead of training them on datasets. When candidates are predicted to perform too poorer than expected, they are immediately killed so that less time will be wasted on those of worse potential.
- Weight inheritance, which reuses weights of previous architectures by initializing newly generated candidates with the same weights as previous similar ones so that there is no need for them to train from scratch. Some works try to inherit weights by first recognizing which parts of candidates are the same as previously trained architectures, while others directly force new candidates to grow in size with previous ones as base structures so that in each search step, only weights of newly added parts need fine-tuning.
- Weight Sharing. This kind of method usually first creates a supergraph, i.e., supernet, and all candidates can be drawn as subgraphs of it, including weights and structures. Since all weights of candidates are shared and updated together, there is no need for separate training of each subgraph. This technique greatly reduces computing time for NAS to only a few GPU days. However, it is concerned that weight sharing puts constraints on candidates and it may incur a large gap of performance estimation between architectures with weight sharing and ones with separate weight training. Some research also points out its inconsistency in ranking architectures [116].

We will next focus on methods of learning curve extrapolation and weight sharing, and skip the other two that are already introduced in detail by previous chapters.

A. Weight Sharing

Weight sharing is proposed in ENAS [93], which is an RL-based NAS method using architecture accuracy to update the controller. To efficiently estimate architecture candidates, ENAS constructs a super network that contains all possible architectures in it, i.e., edges exist between every two nodes and all operation candidates exist on all edges. Each subgraph in the super network represents an architecture. When an architecture needs evaluating, it is not trained, but directly use the weights in the super network. Thus, all the same operations at the same position of all possible architecture share the same weight that is in the super network. E.g., assume A and B are both architectures, both of which choose the edge from node i to node j and choose the operation on this edge as a 3×3 convolution. Then the two convolution layers in both A and B use the weight on the edge from i to j in the super network. To train the super network, the agent should sample some architectures by the controller regularly in the searching phase. Then we adopt forward propagation on these architectures using the weights in the super network. After that, we adopt backpropagation to get the gradients, the weights in the super network will be updated by gradient descent methods using these gradients. The time budget of training a super network is almost the same as training architecture candidates, and an evaluation process of an architecture candidate only costs a little time. Hence, ENAS reduces the running time of searching a neural architecture by an extremely large margin compared with previous works. Since both the efficiency and effectiveness of weight sharing technique, many works design NAS methods base on the super network, named one-shot NAS. One-shot NAS has become an important class in the NAS method family.

   Since the weight sharing technique has become more and more popular, there is a debate on whether the shared weight is accurate enough to give estimated performances. Some works [79, 80] find that the validation accuracy with shared weights is not predictive to its true performance, causing incorrect ranking of architecture candidates, introducing bias and instability to the search process. Bender et al. [100] find that it is possible for one-shot search methods to efficiently identify promising architectures from a complex search space without either hypernetworks or RL. Bender et al. [56] demonstrate that weight sharing could find good architectures in different domains while outperforming the random search method. Deng et al. [68] further explore training/test disparity and mode collapse in the super network, and apply it into several scenarios like model ensemble, uncertainty estimation, and semi-supervised learning.

B. Performance Predictor

Some works train the model for a few epochs to collect data, and use them to predict the whole learning curve so that models predicted to be worse will be early stopped. Previous works take as features points (accuracies or other metrics) collected at initial few epochs and model hyper-parameters to train predictors, like weighted ensemble of parametric functions [69], Bayesian Networks [77] or v-SVR [55].

Other works view only models' final performance as training and testing targets instead of considering the learning curve. In this way, model features are important for prediction. Some of these works first learn to encode networks and then use features obtained to predict final performance by another model, like LSTM [68] or Random Forest [74]. Others exploit neural networks to train in an end-to-end manner. For instance, NAO [89] uses an LSTM to encode model embeddings, followed by an MLP to predict the performance. The model is trained against least-square regression loss targeting at ground-truth performance. The new embedding updated by gradient descent is decoded by a decoder to get the next neural architecture. The authors of [82] propose to search architectures in a heuristic way, i.e., sequential model-based optimization, which progressively expands cells with more blocks, predicts performance, and selects top-k candidates to truly train and evaluate. It uses the ensemble of LSTM and MLP to predict performance for NAS in image tasks. Similarly, the algorithm in [97] iteratively uses the RL controller to generate candidates and select top-k using predictors. It designs a domain-specific language to encode RNN architectures and uses an LSTM as the ranking function to estimate the performance of generated architectures. Another algorithm in [104] uses GCN to encode architectures, followed by an MLP to predict absolute latency or accuracy relationship between two architectures. Accuracy predictions are used to select top-k models to be trained next time, and fully-trained architectures and the corresponding performance are then used to train the predictor. For accuracy prediction, the model takes a pair of architectures as inputs and outputs a two-dimensional vector as probability distribution showing how possibly the first architecture is better than the other. Predicting the binary relationship of architecture performance instead of absolute accuracy is because that the stability and accuracy of ranking candidates are more important than those of predicting absolute accuracy, as the former is in practice what NAS algorithms are utilized to search. Another reason for doing this is that training absolute accuracy predictor is much more challenging and more computationally expensive since more data are needed to be collected for a well-trained absolute predictor.

Supervised methods usually need numbers of labeled data (architecture samples and the corresponding performance) and the collection of these data is computationally expensive. From a semi-supervised perspective, the authors of [102] propose to exploit intrinsic structure similarity and the performance relationship between architectures using a large amount of unlabeled data and only a small amount of labeled data. Specifically, it uses an auto-encoder to obtain architecture embeddings X, which are then utilized to calculate the relationship A between architectures using the Radial Basic Function kernel. Based on the assumption that architectures with similar structures and operations should be similar in real performance, it uses a GCN taking X as node embeddings and A as an adjacency matrix to predict the performance of all architectures. Reconstruction loss and performance regression loss are weighted to train both the autoencoder and GCN. It shows better prediction accuracy and rankings than previous work while using the same amount of labeled data.

To further light the burden on labeled data collection and get rid of bias induced in supervised loss, Yan et al. [112] use a graph autoencoder to learn architecture embeddings in an unsupervised way. Unlike hard-coded encodings such as adjacency matrix and operation list that are discrete, the learned embeddings are continuous and compact in space, while retaining intrinsic relationship between architectures. The embeddings obtained are shown to be helpful for performance predictions, evidenced by results of high correlations between Gaussian process predictions and models' actual performance after full training.

C. Performance Estimation Issues in NAS with Weight Sharing

One of the representative methods of NAS with weight sharing is one-shot NAS, where a supernet is created subsuming all possible candidates and their weights are shared and trained together instead of trained separately. Candidates are evaluated using the shared weights, and in the test stage, the best candidates selected based on evaluation performance are decoded into discrete models and retrained to get test performance. Supernet itself can be viewed as a performance estimator for architectures since the shared weights may not be optimal for each architecture but can represent its potential performance to some extent. To get performance estimation, some works train supernet by relaxing discrete architecture parameters into continuous space and jointly optimizing them with supernet weights. Other works, also called single-path NAS, sample and optimize a single path (an architecture candidate) at a time so that working GPU memory will be saved and it does not bother to transform architecture parameter space from discrete to continuous. This kind of method significantly reduces computational costs and rapidly becomes a hot research topic in the field of NAS.

However, it is observed that the performance between evaluation and test may be inconsistent and the gap may be significant to disturb the effectiveness of architecture search. For single-path one-shot NAS that samples and trains one candidate from supernet at a time like randomNAS and GDAS, the issue is also called the multi-model forgetting problem that when a sequence of models with shared weights are trained using a single dataset, previous models experience performance degradation during training of current models. The algorithm in [119] solves this problem by adding a soft regularization loss to make shared weights of the current model divert less from those of previous representative models. The representatives are selected based on their novelty so that they are diverse enough to cover more submodels. Specifically, novelty is defined as its mean distance from its k-nearest neighbors in the model set, where the distance is calculated based on architecture encodings including edge connections and operations.

Another issue is about the training efficiency of supernet. Many works sample paths by uniform sampling based on the assumption that all paths in supernet should be equally estimated, which includes potentially good as well as poor architectures all the time. However, after training the supernet, only well-performed architectures are useful, so actually, a large amount of time is wasted for training poor architectures and their training may also harm good architectures since their weights are highly shared with better ones. To this end, the authors of [115] propose a multi-path sampling strategy to greedily train potentially good paths and filter poor ones. Concretely, it uses a pool to store top-k performed paths, and it uses uniform sampling in both the pool maintained and the whole search space. The sample number from both sources is assigned by a scheduler, which gradually increases the ratio of samples from pool to shift training from exploration to exploitation. In this way, it is faster and better to estimate the performance of potentially good paths, which are our main concern.

For general NAS, a variety of previous works have been dedicated to accelerating performance estimation of architectures by budgeting computational resources, namely Budgeted Performance Estimation (BPE). It is usually done by constraining algorithm hyper-parameters to a smaller scale, e.g., utilizing a proxy model with a smaller size, sampling a smaller but property-preserving dataset, early stopping for training, etc. However, most of those methods are hand-designed intuitively that may not guarantee the consistency between the BPE and final full evaluation of architectures. To achieve both acceleration and consistency of BPE, the authors of [122] summarize several hyper-parameter reduction methods usually used in previous works and find the Pareto front of these methods by experimenting on a set of factor combinations. The hyper-parameters for variation include the number of channels for CNNs, resolution factor of input images, the number of training epochs, and sample ratio of the full training dataset. They discover that increasing training samples with fewer training epochs is more consistent than using a combination with fewer samples and more epochs. Another discovery about the resolution of input images and channels of networks also shows that different hyper-parameters contribute differently to the properties of BPE. Based on those discoveries, the method applies the most efficient BPE proxy found to Amoeba-Net and improves nearly 400x the training speed. Instead of trying and evaluating different combinations of factors, the authors of [120] proposes to model the relationship between hyper-parameters and the corresponding estimation properties, where acceleration is evaluated by average FLOPS for training models and consistency is evaluated by Spearman Rank Correlation between models trained with full hyper-parameters and ones trained with budgeted hyper-parameters. Note that here hyper-parameter is to be searched and be evaluated by the average performance of models trained using the selected hyper-parameter. Concretely, hyper-parameters are sampled with criteria that ones with lower time cost are more likely to be sampled. Then a Random Forest model will predict their properties including acceleration and consistency as mentioned above. Hyper-parameters with the lowest properties are assigned with minimum computational resources so that those of low importance will be pruned. A set of models will be trained for each hyper-parameter with corresponding computational resources, and their performance will be used to update the Random Forest. Finally, the optimal hyper-parameter obtained can act as an efficient proxy for NAS. As reported, GPU hours needed for searching optimal hyper-parameter can be reduced 1000 times from 7×10^4 to 5.2×10^3.

Some works pursue sufficient training of supernet including a variety of potentially good subnetworks so that the supernet can give more accurate performance estimation. The authors of [92] propose to select well-performed paths as teachers on the fly during training and use them to guide the training of current paths. Concretely, paths with better accuracy and lower FLOPS are appended into a fixed-size priority queue, and worse paths are knocked out if the queue is full. When training other paths, a teacher network, also termed as the prioritized path, is drawn from the queue by a meta-network to guide training by knowledge distillation. The meta-networks learn to find the best-matched prioritized path with the current path by considering the difference of their feature logits so that distinct parts of teacher networks can be useful as complementary information for training current paths. The algorithm first initializes weights of supernet and priority queue, followed by an interleaving of random sampling paths, choosing prioritized paths, path training with distillation, and meta-networks updating. At last, the best solution can be directly obtained from the queue maintained instead of sampling or using other selection methods.

1.3 Advances on AutoML

In this section, we will introduce the advances on AutoML, including grey-box AutoML, hardware-aware AutoML, AutoML for Reinforcement Learning and AutoML for graph. These topics are emerging directions in AutoML research and we believe they may inspire the readers for future investigations.

1.3.1 Grey-Box AutoML

In black-box HPO methods using the sequential model-based optimization framework (see Algorithm 1), the performance evaluation can be very expensive, especially for large datasets and complex models. Specifically, each time some certain configuration is evaluated, we need to run a complete training procedure of the machine learning system being optimized, and then measure the performance of the derived model on the full validation set. Both model training and loss validation can be expensive, which may cause a waste of the computational resource. Recently, a surge of researches have focused on lowering down the evaluation expense by utilizing some side information collected from the evaluation procedure. With these approaches, the problem of HPO essentially becomes a "grey-box", namely, although we may not have access to the full curvature of the target function as in the white-box problem, we do know something more than the black-box optimization problem where the only information is the function value at the evaluation point. In the following, we will introduce three main branches of grey-box AutoML methods, which are designed to accelerate the evaluation of candidate configurations.

A. Hyper-Parameter Gradient Descent

Recall that, the objective of HPO takes

$$\lambda^* = \arg\min_{\lambda \in \Lambda} \mathcal{V}(A_\lambda, \mathcal{D}).$$

From a higher (or *meta*) point of view, the hyper-parameters in algorithm configuration can be viewed as learnable parameters, which can then be optimized using the techniques in conventional optimization problems [14, 15]. We here introduce approaches that directly optimize λ in such a way, namely, through its gradients.

Given the datasets for training $\mathcal{D}_{train}$ and validation $\mathcal{D}_{val}$, [15] formulates HPO as a bilevel optimization problem, which is given as

$$\lambda^* = \arg\min_{\lambda \in \Lambda} \mathcal{L}(\phi_\lambda, \mathcal{D}_{val}), \quad s.t. \quad \phi_\lambda = \arg\min_{\phi \in \Phi} \mathcal{L}(\phi, \mathcal{D}_{train}).$$

In the above definition, ϕ represents the model parameters from the parameter space Φ, and $\mathcal{L}(\phi, \mathcal{D})$ measures the loss of the model parameterized by ϕ on the dataset $\mathcal{D}$. The core step in the bi-level optimization is to calculate the gradient w.r.t. the hyper-parameters λ (also called the hypergradient). Several methods have been proposed for efficient hypergradient calculation. Among these approaches, [34] calculates the hypergradient by reversing the dynamics of stochastic gradient descent with momentum. Reference [33] tunes continuous hyper-parameters using the gradient of the performance of the model on a separate validation set. When the hyper-parameters λ is real-valued, [15] discusses a general approach for efficient hypergradient calculation by using a dynamical system to approximate the optimization function and then applying algorithmic differentiation [2], which enjoys computational efficiency as well as provable convergence guarantees.

B. Extrapolation of Learning Curves

In the basic SMBO framework, for each evaluation, we need to experience the full procedure of model training in order to observe the final validation performance. However, for some undesirable configurations, we do not really need to run the full training procedure, but can terminate the evaluation based on their poor performance in the early stages of model training. In this spirit, [9] proposes to extrapolate the final performance by fitting a probabilistic model (termed the parametric learning curve model) to the learning curve at the early stages, and then terminate the evaluation on configurations which are predicted to have poor final performance. Their approach is proved suitable to HPO of deep neural networks, where the training and evaluation of deep models are very expensive.

C. Multi-Fidelity Optimization

Since the evaluation of the black-box function is expensive for itself, we can use cheap approximations of the black-box functions to evaluate the performance of the candidate configurations. A wide range of methods have been developed in this philosophy, which are also called the multi-fidelity methods. In the previous part, we have introduced a major branch of multi-fidelity methods, namely, the bandit-based methods (e.g., Successive Halving [26] and Hyperband [32]), which run algorithms with multiple configurations at the same time and then rule out those with relatively bad performance in the early stages. Here we describe another branch of multi-fidelity methods, which realizes cheap evaluation by training the model with given candidate configurations on a small amount of data and/or with a small number of epochs [27, 28, 51]. These methods fit a Gaussian process $M(\lambda, b)$ to predict the validation performance as a function of configuration λ as well as budget b (which can be either the amount d of training data or the number t of training epochs), such that $\lim_b M(\lambda, b) = M(\lambda) \approx M(\lambda, B)$ where B is the total budget. To leverage the budget consumption, the corresponding acquisition function α is also tweaked by a cost function c, which now takes $\alpha(\lambda, b)/c(\lambda, b)$ [27].

1.3.2   Hardware-Aware AutoML

Deep neural networks (DNNs) have achieved great success nowadays. The performance of neural networks heavily relies on their architectures. However, the manual design is usually tedious and time-consuming, motivating the advent of Neural Architecture Search (NAS) which automatically searches for the optimal architecture. As increasing devices are equipped with DNNs, there are higher requirement to design more efficient architectures for vast amount of hardware, which is quite challenging due to the complexity of the various hardware designs. Besides, most existing research on NAS use hardware-agnostic metrics such as FLOPs to measure the model efficiency or estimate the inference latency, causing the inconsistency between these hardware-agnostic metrics and the actual efficiency on the real hardware.

However, with the complexity of tasks and huge data, obtaining a model with high efficiency in hardware platform become more and more necessary. At this time, FLOPs began to be added into the training process to obtain a model with lower FLOPs. However, some researchers [141–144, 147, 148] have come to realize that the relationship between FLOPs and the real performance of the model on specific hardware is inaccurate. Therefore, Hardware-aware performance metrics (latency, power, memory etc.) are added into the search process instead of FLOPs to measure the efficiency of the model better. At present, an objective of researchers is to reduce the latency of the model which is found by NAS.

In order to achieve the purpose of both high accuracy and efficiency on specific hardware platform, a variety of methods are used to combine the two objectives. For example, [144] defines this multi-objective problem as follows:

Give a model m, let $ACC(m)$ denote its accuracy on the target task, $LAT(m)$ denotes the inference latency on the target hardware platform, and T is the target latency. One kind of the optimization goal can be defined as:

$$\underset{m}{\text{maximize}} \quad ACC(m) \times \left[\frac{LAT(m)}{T} \right]^{w}.$$

(1.37)

In this way, we can get a batch of excellent models, which have good performance in both accuracy and efficiency.

A. Hardware-Aware Search Space

The search space specifies which kinds of neural architectures can be discovered in principle. As the optimization problem of NAS is often non-continuous and high-dimensional, it can be largely simplified by a suitable choice of the search space, since we can reduce the size of the search space by incorporating prior knowledge on the specific task.

Neural architecture search usually aims to automatically discover the optimal topology structure, while the depth and width of the backbone architecture are fixed. However, the width and depth can also significantly influence the performance of the model. The neural network with a wider width and deeper depth usually has higher accuracy but also more computation cost. Therefore, to achieve a balance between the model efficiency and the prediction accuracy or adapt to different tasks, Transformable Architecture Search (TAS) [149] adjusts the network size, through searching for the best width and depth configuration for the network backbone.

B. Hardware-Aware Performance Metrics

One of the feasible ways to obtain hardware-aware metrics is to measure them directly in the search process [141]. But for a NAS framework that updates both model parameters and search parameters at the same time, a large number of measurement will lead to a large extension of the search process.

As we know, a large number of search process can only be trained on large GPUs, but with the wide use of mobile devices, more and more models obtained are to run on the mobile device. Moreover, considering that the metrics on actual hardware are affected by many factors, the metrics of model are not stable if we measure them during the search process, and the search process will be greatly prolonged if averaging many measurements to gain metrics.

For the above reasons, researchers use various methods to obtain stable and reliable metrics of model on specific hardware platform. For example, [145] builds a huge database which includes the true metrics of all possible models on specific hardware platform, [147] assumes that the latency of model can be represented by the latency of each layer to reduce the amount of metrics to be measured. In addition, a more common method is to use neural network model to predict metrics, [146] uses MLP, [140] uses GCN, and so on.

C. Hardware-Aware Search Method

How to explore the search space is a very important topic. The most direct and effective way is to add them to loss function. For a multi-objective problem, it's obvious that a solution can not make all the objectives optimal. Instead, the search process is expected to produce solutions satisfying Pareto Optimality. Pareto optimal solution means there is no solution that is superior to it in at least one objective and not inferior to it in all objectives.

By designing the loss function reasonably, it is possible to obtain multiple Pareto optimal solutions. References [139, 142] use weighted sum of latency and other terms as the loss function. References [144, 147] use weighted product of latency and other terms.

Besides, Hardware-aware performance metrics can be used to do more. For example, Jiang et al. [142] use metrics to filter the search space, which makes the model obtained has excellent performance on specific hardware platform.

1.3.3   AutoML for Reinforcement Learning

Reinforcement learning (RL) [135] is an important learning paradigm which aims to maximize the cumulative rewards an agent obtains from interacting with an environment. The key of reinforcement learning is to learn an optimal policy $\pi(a|s)$, i.e., a probability distribution over the action space conditioned on the current state, to maximize the agent's long-term returns.

AutoML for reinforcement learning is not a simple application of AutoML to a new learning scenario, but imposes several new challenges due to the characteristics of reinforcement learning. First, reinforcement learning is usually more time-consuming to train compared to supervised learning, especially when the interaction samples need to be collected by interacting with the physical world [130], which makes computational efficiency and sample efficiency a critical issue in AutoML for reinforcement learning. Second, the learning process of reinforcement learning is very unstable compared to supervised learning due to its non-stationary data distribution and learning techniques like bootstrapping, which makes it very sensitive to the hyper-parameters [129]. Third, even under the same hyper-parameter setting, significant performance variance has also been observed between different trials due to intrinsic randomness in reinforcement learning [129] (such as simply changing the random seed), which makes it much harder to make a fair and accurate comparison between different algorithm configurations.

Different works have been proposed to tackle the above challenges in AutoML for reinforcement learning, and we give an overview of them according to the different components they focus on in the reinforcement learning framework.

Reward Design Reward design is particularly important in reinforcement learning, as the reward defines the task objective and desirable behaviors of the agent. Designing a proper reward signal is a difficult and tedious task which requires many iterations of manual tuning based on domain knowledge. Some previous works aim to tackle this problem by automatically designing a good reward signal. AutoRL [126, 127] learns the reward shaping function with evolutionary algorithm, which shows improvement in both robot navigation and continuous control tasks. Reference [125] focuses on designing the curiosity algorithms (a kind of intrinsic reward) [134] in reinforcement learning. It formulates different formulations of curiosity as different architectures in a search space defined by a domain-specific language. It also uses some AutoML techniques to accelerate the searching process, such as pruning less promising candidates in cheap environments and performance prediction based on meta features. The learned curiosity algorithms not only outperform human-designed algorithms in the training environments, but also generalize well to even significantly different and more complex environments.

HPO and Algorithm Design A variety of reinforcement learning algorithms have been proposed in the past decades, and the performance are very sensitive to their hyper-parameters [129]. Most previous works on AutoML for reinforcement learning tune these hyper-parameters on-the-fly, i.e., the hyper-parameters are adaptively tuned in each iteration during the training process, instead of treated as an outer-loop learning objective. The motivations of this on-the-fly setting are mainly twofold. First, reinforcement learning is usually very time-consuming, thus evaluating all different hyper-parameter configurations is usually infeasible. Second, the data distribution encountered by the agent is non-stationary and changes with the agent's policy, thus using a fixed hyper-parameter configuration along the whole learning process usually leads to sub-optimal performance. Different methods have been proposed for on-the-fly HPO in reinforcement learning. Meta-gradient [137, 138] methods compute the gradients of the hyper-parameters w.r.t. the learning objective and update them accordingly. However, these methods are limited to the hyper-parameters with explicit gradients. HOOF (Hyper-parameter Optimization on the Fly) [132] optimizes the hyper-parameters with random search, and utilizes importance sampling to speed up the evaluation of each configuration. Reference [136] uses evolutionary strategies to optimize the hyper-parameters in off-policy learning algorithms. While the above mentioned works focus on HPO of existing reinforcement learning algorithms, LPG (Learned Policy Gradient) [131] takes a step further by directly learning the update rule in reinforcement learning algorithms with meta-gradient. It automatically discovers its own alternatives to the basic concepts and mechanisms in reinforcement learning such as value function and bootstrapping, and the learned algorithms can effectively generalize to more complex environments than those used for algorithm learning.

Neural Architecture Search In the past decade, deep learning models have been widely adopted in reinforcement learning as the function approximator for different components in the framework. Some previous works have also investigated how to find the optimal structures of these networks in reinforcement learning based on evolutionary algorithms [126], Bayesian optimization [133], and on-the-fly neuro-evolution [128]. However, these works mainly focus on some very simple architectures like MLPs, and it remains to explore more efficient NAS algorithms for more complex network architectures in reinforcement learning in future work.

1.3.4 AutoML for Graph

Graph data is ubiquitous in representing relational data such as social networks, knowledge graphs, biomedical networks, the World Wide Web, etc. Typically, graphs are composed of nodes representing objects and edges representing relationships between objects. Due to such complicated dependencies, graph data poses unique challenges for machine learning models [151, 154]. For example, it is non-trivial to split a large-scale graph into small independent batches since the nodes and edges are correlated. Thus, how to deal with graph data in AutoML is a challenging issue.

A. Hyper-Parameter Optimization for Graph

We introduce AutoNE [153], an HPO method for machine learning models on graphs such as network embedding [151] or graph neural networks (GNNs). As explained above, one key difficulty of HPO for graphs lies in that we cannot trivially separate the graph into different components for batch learning, while directly optimizing the hyper-parameters on the whole graph is computationally expansive, especially for large-scale graphs such as social networks or the Web with millions or billions of nodes and edges. To tackle this issue, one solution is to sample the graph, conduct HPO on the sampled subgraphs, and then transfer the hyper-parameters back into the original graph. The remaining questions are three-folded. First, how can we effectively sample the graph. Second, how to conduct HPO on the sampled subgraphs. Lastly, how to transfer the hyper-parameters.

The sampling problem is non-trivial because a graph can consist of multiple heterogeneous components. Without careful designs, important information may be lost in the sampling process. AutoNE proposes a multi-start random walk strategy, i.e., multiple starting points of random walks are randomly chosen to preserve the heterogeneity of the graph structure. Specifically, if node labels are available, starting points are evenly distributed among different labels. If no node label is available, FastGreedy [150], a greedy community detection algorithm is applied so that the starting points are evenly distributed among different communities.

After obtaining the sampled subgraphs, we can conduct HPO on these subgraphs. To enable knowledge transfer among different subgraphs, we need to calculate and compare the similarities between different graphs. Thus, we resort to NetLSD [152], an unsupervised graph representation learning algorithm as the graph feature extractor, and adopt the Matern 5/2 kernel function [45] to calculate similarities. Compared with other kernels such as the radial basis function (RBF) kernel, the Matern 5/2 kernel is more capable of describing non-smooth functions. Then, we can use the normal Bayesian optimization with the Gaussian process to optimize the hyper-parameters on sampled subgraphs.

The final step is to optimize the hyper-parameters on the original graph. Notice that by using NetLSD, which is size-invariant, and the kernel functions, we can estimate the similarities of the original graph with the sampled subgraphs. Thus, we can directly feed the feature of the original graph into the Gaussian process to predict the optimal hyper-parameters. However, notice that though HPO on sampled subgraphs provides helpful information in learning optimal hyper-parameters on the original graph, AutoNE experimentally shows that a few trials on the original graph are still needed to "calibrate" the HPO model to achieve promising performance.

B. Neural Architecture Search for Graph

Graph-structured data such as social network and biological molecules has attracted lots of attentions in recent years. Manually designed graph neural networks (GNNs) has been widely applied on graph structured data and achieved successful results. However, the design process needs lots of human efforts and domain knowledge. Recently, some papers apply NAS methods on designing GNNs.

Automatically designed GNNs are applied on semi-supervised node classification tasks at the very beginning. GraphNAS [155] and AutoGNN [161] are the earliest attempts on using NAS methods to designs GNNs. Both of them use RL-based NAS methods. They decompose a GNN layer into actions following.

- Hidden dimension: the dimension of embedding in current GNN layer.
- Sample function: the receptive field of nodes in the GNN layer.
- Attention function: the mechanism of assigning importance of neighbors.
- Attention head: the number of multiple attention heads.
- Aggregation function: the mechanism of aggregating data from neighbors.
- Combine function: the mechanism of combining information of node itself and neighbors.
- Residual function: the mechanism of merging previous layers' representation.
- Activation function: the non-linear function adopted to the layer output.

These actions are adopted one by one to data. Table 1.1 shows the designed actions in different methods. GraphNAS utilizes an RL controller to generate architectures and uses policy gradient to train it. For evaluation, GraphNAS uses weight sharing

Table 1.1 Decomposition of GNN layer

Action	GraphNAS	AutoGNN	PDNAS	NASGNN
Hidden dimension	✓	✓	✓	✓
Sample function	✓			
Attention function	✓	✓	✓	✓
Attention head		✓	✓	✓
Aggregate function	✓	✓	✓	✓
Combine function		✓	✓	
Residual function	✓			
Activation function	✓	✓	✓	✓

and allows different attention functions share one transform weights. However, this strategy may cause deviation, so GraphNAS fine tunes the child model with shared parameters to obtain more precise reward. Finally, GraphNAS chooses the architecture with maximum predicted reward. AutoGNN maintains the best architecture during searching. In each turn, AutoGNN uses RNN encoders and action guider to determine how to modify the best architecture. If the modified architecture has the same dimension, attention function and activation function, it will directly uses the best architecture's weight, otherwise the modified architecture is trained. Then the performance is used to update RNN encoders and action guider by policy gradient.

PDNAS [160] applies gradient-based NAS methods on designing GNNs. PDNAS search in both micro and macro space. Micro space searches the actions inside a layer, as shown in Table 1.1, while macro space searches the connections among different layers. For micro space, PDNAS uses an MLP to generate architecture from a trainable prior vector. For macro space, PDNAS uses Gumbel distribution to decide the existence of each possible edges. The training procedure follows a dual optimization scheme.

Nunes et al. [158] applies Aging Evolution algorithm on NAS to design GNNs. They find that both RL based and evolution based NAS methods obtain similar accuracies to a random search, raising the question of how many of the search space dimensions are actually relevant to the problem.

Besides node classification, automatically designed GNNs are also applied on other tasks. For link prediction task, NASE [156] decomposes the architecture into representation search module and score function search module. NASE provides several candidates for both modules, and constructs an end-to-end super network. Therefore, NASE can use DARTS to optimize the architecture.

For skeleton-based human action recognition task, CEIM [159] constructs the search space with structure representation convolution and temporal representation convolution to capture information of human skeleton. Besides, CEIM builds Chebyshev polynomials functions with different order at different layers to automatically choose the components. For the search strategy, CEIM models the architecture as Gaussian distribution. In each epoch, CEIM samples architectures from the distribution, then selects part of them to do evaluation. Importance weights are assigned to those architectures according to their performance. Then the parameters of Gaussian distribution are updated with those weighted architectures.

More recently, SGAS [79] applies designed GNN on 3D object classification dataset ModelNet and biological dataset PPI. SGAS only searches for the macro space of GNN, that means the micro space of layers are predetermined. SGAS provides 10 candidate layers. During search procedure, these layers are acted as operations as in CNN searching. This search space enables the direct applications of all existing NAS methods on GNN searching tasks.

References

1. Mohamed Osama Ahmed, Bobak Shahriari, and Mark Schmidt. Do we need "harmless" bayesian optimization and "first-order" bayesian optimization. In *NeurIPS Workshop on Bayesian Optimization (BayesOpt'16)*, 2016.
2. Atilim Gunes Baydin, Barak A Pearlmutter, Alexey Andreyevich Radul, and Jeffrey Mark Siskind. Automatic differentiation in machine learning: a survey. *Journal of machine learning research*, 18, 2018.
3. James Bergstra and Yoshua Bengio. Random search for hyper-parameter optimization. *Journal of machine learning research*, 13(2), 2012.
4. James Bergstra, Rémi Bardenet, Yoshua Bengio, and Balázs Kégl. Algorithms for hyper-parameter optimization. In *25th annual conference on neural information processing systems (NIPS 2011)*, volume 24. Neural Information Processing Systems Foundation, 2011.
5. James Bergstra, Daniel Yamins, and David Cox. Making a science of model search: Hyper-parameter optimization in hundreds of dimensions for vision architectures. In *International conference on machine learning*, pages 115–123. PMLR, 2013.
6. Bernd Bischl, Olaf Mersmann, Heike Trautmann, and Claus Weihs. Resampling methods for meta-model validation with recommendations for evolutionary computation. *Evolutionary computation*, 20(2):249–275, 2012.
7. Eric Brochu, Vlad M Cora, and Nando De Freitas. A tutorial on bayesian optimization of expensive cost functions, with application to active user modeling and hierarchical reinforcement learning. *arXiv preprint arXiv:1012.2599*, 2010.
8. Bo Chen, Rui Castro, and Andreas Krause. Joint optimization and variable selection of high-dimensional gaussian processes. In *Proceedings of the 29th International Conference on Machine Learning.*, pages 1423–1430. International Machine Learning Society, 2012.
9. Tobias Domhan, Jost Tobias Springenberg, and Frank Hutter. Speeding up automatic hyperparameter optimization of deep neural networks by extrapolation of learning curves. In *Twenty-fourth international joint conference on artificial intelligence*, 2015.
10. Russell C Eberhart and Yuhui Shi. Comparison between genetic algorithms and particle swarm optimization. In *International conference on evolutionary programming*, pages 611–616. Springer, 1998.
11. Hugo Jair Escalante, Manuel Montes, and Luis Enrique Sucar. Particle swarm model selection. *Journal of Machine Learning Research*, 10(2), 2009.
12. Stefan Falkner, Aaron Klein, and Frank Hutter. Bohb: Robust and efficient hyperparameter optimization at scale. In *International Conference on Machine Learning*, pages 1437–1446. PMLR, 2018.
13. Matthias Feurer, Aaron Klein, Katharina Eggensperger, Jost Tobias Springenberg, Manuel Blum, and Frank Hutter. Auto-sklearn: efficient and robust automated machine learning. In *Automated Machine Learning*, pages 113–134. Springer, Cham, 2019.

14. Luca Franceschi, Michele Donini, Paolo Frasconi, and Massimiliano Pontil. Forward and reverse gradient-based hyperparameter optimization. In *International Conference on Machine Learning*, pages 1165–1173. PMLR, 2017.

15. Luca Franceschi, Paolo Frasconi, Saverio Salzo, Riccardo Grazzi, and Massimiliano Pontil. Bilevel programming for hyperparameter optimization and meta-learning. In *International Conference on Machine Learning*, pages 1568–1577. PMLR, 2018.

16. Nikolaus Hansen. The cma evolution strategy: a comparing review. *Towards a new evolutionary computation*, pages 75–102, 2006.

17. Nikolaus Hansen. The cma evolution strategy: A tutorial. *arXiv preprint arXiv:1604.00772*, 2016.

18. Kaiming He, Xiangyu Zhang, Shaoqing Ren, and Jian Sun. Deep residual learning for image recognition. In *Proceedings of the IEEE conference on computer vision and pattern recognition*, pages 770–778, 2016.

19. Philipp Hennig and Christian J Schuler. Entropy search for information-efficient global optimization. *Journal of Machine Learning Research*, 13(6), 2012.

20. José Miguel Hernández-Lobato, Matthew W Hoffman, and Zoubin Ghahramani. Predictive entropy search for efficient global optimization of black-box functions. *arXiv preprint arXiv:1406.2541*, 2014.

21. Matthew D Hoffman, Eric Brochu, and Nando de Freitas. Portfolio allocation for bayesian optimization. In *UAI*, pages 327–336. Citeseer, 2011.

22. Frank Hutter. *Automated configuration of algorithms for solving hard computational problems*. PhD thesis, University of British Columbia, 2009.

23. Frank Hutter, Holger H Hoos, and Kevin Leyton-Brown. Sequential model-based optimization for general algorithm configuration. In *International conference on learning and intelligent optimization*, pages 507–523. Springer, 2011.

24. Frank Hutter, Holger H Hoos, and Kevin Leyton-Brown. Identifying key algorithm parameters and instance features using forward selection. In *International Conference on Learning and Intelligent Optimization*, pages 364–381. Springer, 2013.

25. Frank Hutter, Lars Kotthoff, and Joaquin Vanschoren. *Automated machine learning: methods, systems, challenges*. Springer Nature, 2019.

26. Kevin Jamieson and Ameet Talwalkar. Non-stochastic best arm identification and hyperparameter optimization. In *Artificial Intelligence and Statistics*, pages 240–248. PMLR, 2016.

27. Kirthevasan Kandasamy, Gautam Dasarathy, Jeff Schneider, and Barnabás Póczos. Multi-fidelity bayesian optimisation with continuous approximations. In *International Conference on Machine Learning*, pages 1799–1808. PMLR, 2017.

28. Aaron Klein, Stefan Falkner, Simon Bartels, Philipp Hennig, and Frank Hutter. Fast bayesian optimization of machine learning hyperparameters on large datasets. In *Artificial Intelligence and Statistics*, pages 528–536. PMLR, 2017.

29. Ron Kohavi and George H John. Automatic parameter selection by minimizing estimated error. In *Machine Learning Proceedings 1995*, pages 304–312. Elsevier, 1995.

30. Brent Komer, James Bergstra, and Chris Eliasmith. Hyperopt-sklearn. In *Automated Machine Learning*, pages 97–111. Springer, Cham, 2019.

31. Miguel Lázaro-Gredilla, Joaquin Quinonero-Candela, Carl Edward Rasmussen, and Aníbal R Figueiras-Vidal. Sparse spectrum gaussian process regression. *The Journal of Machine Learning Research*, 11:1865–1881, 2010.

32. Lisha Li, Kevin Jamieson, Giulia DeSalvo, Afshin Rostamizadeh, and Ameet Talwalkar. Hyperband: A novel bandit-based approach to hyperparameter optimization. *The Journal of Machine Learning Research*, 18(1):6765–6816, 2017.

33. Jelena Luketina, Mathias Berglund, Klaus Greff, and Tapani Raiko. Scalable gradient-based tuning of continuous regularization hyperparameters. In *International conference on machine learning*, pages 2952–2960. PMLR, 2016.

34. Dougal Maclaurin, David Duvenaud, and Ryan Adams. Gradient-based hyperparameter optimization through reversible learning. In *International conference on machine learning*, pages 2113–2122. PMLR, 2015.

35. Mitchell McIntire, Daniel Ratner, and Stefano Ermon. Sparse gaussian processes for bayesian optimization. In *UAI*, 2016.
36. Gábor Melis, Chris Dyer, and Phil Blunsom. On the state of the art of evaluation in neural language models. In *International Conference on Learning Representations*, 2018.
37. Volodymyr Mnih, Csaba Szepesvári, and Jean-Yves Audibert. Empirical bernstein stopping. In *Proceedings of the 25th international conference on Machine learning*, pages 672–679, 2008.
38. Jonas Mockus, Vytautas Tiesis, and Antanas Zilinskas. The application of bayesian methods for seeking the extremum. *Towards global optimization*, 2(117–129):2, 1978.
39. Douglas C Montgomery. *Design and analysis of experiments*. John wiley & sons, 2017.
40. Carl Edward Rasmussen. Gaussian processes in machine learning. In *Summer school on machine learning*, pages 63–71. Springer, 2003.
41. Matthias Seeger, Christopher KI Williams, and Neil D Lawrence. Fast forward selection to speed up sparse gaussian process regression. In *In workshop on AI and statistics*, 2003.
42. Bobak Shahriari, Kevin Swersky, Ziyu Wang, Ryan P Adams, and Nando De Freitas. Taking the human out of the loop: A review of bayesian optimization. *Proceedings of the IEEE*, 104 (1):148–175, 2015.
43. Dan Simon. *Evolutionary optimization algorithms*. John Wiley & Sons, 2013.
44. Edward Snelson and Zoubin Ghahramani. Sparse gaussian processes using pseudo-inputs. *Advances in neural information processing systems*, 18:1257–1264, 2005.
45. Jasper Snoek, Hugo Larochelle, and Ryan Prescott Adams. Practical bayesian optimization of machine learning algorithms. *Advances in Neural Information Processing Systems*, 2012.
46. Jasper Snoek, Oren Rippel, Kevin Swersky, Ryan Kiros, Nadathur Satish, Narayanan Sundaram, Mostofa Patwary, Mr Prabhat, and Ryan Adams. Scalable bayesian optimization using deep neural networks. In *International conference on machine learning*, pages 2171–2180. PMLR, 2015.
47. Jost Tobias Springenberg, Aaron Klein, Stefan Falkner, and Frank Hutter. Bayesian optimization with robust bayesian neural networks. In *Proceedings of the 30th International Conference on Neural Information Processing Systems*, pages 4141–4149, 2016.
48. Niranjan Srinivas, Andreas Krause, Sham Kakade, and Matthias Seeger. Gaussian process optimization in the bandit setting: No regret and experimental design. In *Proceedings of the 27th International Conference on Machine Learning*, number CONF. Omnipress, 2010.
49. Chris Thornton, Frank Hutter, Holger H Hoos, and Kevin Leyton-Brown. Auto-weka: Combined selection and hyperparameter optimization of classification algorithms. In *Proceedings of the 19th ACM SIGKDD international conference on Knowledge discovery and data mining*, pages 847–855, 2013.
50. Zi Wang, Clement Gehring, Pushmeet Kohli, and Stefanie Jegelka. Batched large-scale bayesian optimization in high-dimensional spaces. In *International Conference on Artificial Intelligence and Statistics*, pages 745–754. PMLR, 2018.
51. Jian Wu, Saul Toscano-Palmerin, Peter I Frazier, and Andrew Gordon Wilson. Practical multi-fidelity bayesian optimization for hyperparameter tuning. In *Uncertainty in Artificial Intelligence*, pages 788–798. PMLR, 2020.
52. Bubeck, Sébastien, Munos, Rémi, and Stoltz, Gilles. Pure exploration in multi-armed bandits problems. In *International conference on Algorithmic learning theory*, pp. 23–37. Springer, 2009.
53. Li, Lisha, Jamieson, Kevin, DeSalvo, Giulia, Rostamizadeh, Afshin, and Talwalkar, Ameet. Hyperband: A novel bandit-based approach to hyperparameter optimization. *arXiv preprint arXiv:1603.06560*, 2016.
54. Bowen Baker, Otkrist Gupta, Nikhil Naik, and Ramesh Raskar. Designing neural network architectures using reinforcement learning, 2017.
55. Bowen Baker, Otkrist Gupta, Ramesh Raskar, and Nikhil Naik. Accelerating neural architecture search using performance prediction. In *6th International Conference on Learning Representations, ICLR 2018, Vancouver, BC, Canada, April 30 - May 3, 2018, Workshop Track Proceedings*. OpenReview.net, 2018.

56. Gabriel Bender, Hanxiao Liu, Bo Chen, Grace Chu, Shuyang Cheng, Pieter-Jan Kindermans, and Quoc V Le. Can weight sharing outperform random architecture search? an investigation with tunas. In *Proceedings of the IEEE/CVF Conference on Computer Vision and Pattern Recognition*, pages 14323–14332, 2020.

57. James Bergstra and Yoshua Bengio. Random search for hyper-parameter optimization. *Journal of machine learning research*, 13(2), 2012.

58. Han Cai, Tianyao Chen, Weinan Zhang, Yong Yu, and Jun Wang. Efficient architecture search by network transformation. In *Proceedings of the AAAI Conference on Artificial Intelligence*, volume 32, 2018.

59. Han Cai, Jiacheng Yang, Weinan Zhang, Song Han, and Yong Yu. Path-level network transformation for efficient architecture search. In *International Conference on Machine Learning*, pages 678–687. PMLR, 2018.

60. Han Cai, Ligeng Zhu, and Song Han. Proxylessnas: Direct neural architecture search on target task and hardware. In *7th International Conference on Learning Representations, ICLR 2019, New Orleans, LA, USA, May 6–9, 2019*. OpenReview.net, 2019.

61. Jianlong Chang, Yiwen Guo, GAOFENG MENG, SHIMING XIANG, Chunhong Pan, et al. Data: Differentiable architecture approximation. *Advances in Neural Information Processing Systems*, 32:876–886, 2019.

62. Thomas Chau, Łukasz Dudziak, Mohamed S Abdelfattah, Royson Lee, Hyeji Kim, and Nicholas D Lane. Brp-nas: Prediction-based nas using gcns. *Advances in Neural Information Processing Systems*, 33, 2020.

63. Xin Chen, Lingxi Xie, Jun Wu, and Qi Tian. Progressive differentiable architecture search: Bridging the depth gap between search and evaluation. In *Proceedings of the IEEE/CVF International Conference on Computer Vision*, pages 1294–1303, 2019.

64. Yukang Chen, Gaofeng Meng, Qian Zhang, Shiming Xiang, Chang Huang, Lisen Mu, and Xinggang Wang. Reinforced evolutionary neural architecture search. *arXiv preprint arXiv:1808.00193*, 2018.

65. Philip Colangelo, Oren Segal, Alex Speicher, and Martin Margala. Artificial neural network and accelerator co-design using evolutionary algorithms. In *2019 IEEE High Performance Extreme Computing Conference (HPEC)*, pages 1–8. IEEE, 2019.

66. Philip Colangelo, Oren Segal, Alexander Speicher, and Martin Margala. Evolutionary cell aided design for neural network architectures. *arXiv preprint arXiv:1903.02130*, 2019.

67. Jiequan Cui, Pengguang Chen, Ruiyu Li, Shu Liu, Xiaoyong Shen, and Jiaya Jia. Fast and practical neural architecture search. In *2019 IEEE/CVF International Conference on Computer Vision, ICCV 2019, Seoul, Korea (South), October 27 - November 2, 2019*, pages 6508–6517. IEEE, 2019.

68. Boyang Deng, Junjie Yan, and Dahua Lin. Peephole: Predicting network performance before training. *CoRR*, abs/1712.03351, 2017.

69. Tobias Domhan, Jost Tobias Springenberg, and Frank Hutter. Speeding up automatic hyperparameter optimization of deep neural networks by extrapolation of learning curves. In *IJCAI*, pages 3460–3468, 2015.

70. Xuanyi Dong and Yi Yang. Searching for a robust neural architecture in four gpu hours. In *Proceedings of the IEEE/CVF Conference on Computer Vision and Pattern Recognition*, pages 1761–1770, 2019.

71. Thomas Elsken, Jan Hendrik Metzen, and Frank Hutter. Efficient multi-objective neural architecture search via lamarckian evolution. *arXiv preprint arXiv:1804.09081*, 2018.

72. Chaoyang He, Haishan Ye, Li Shen, and Tong Zhang. Milenas: Efficient neural architecture search via mixed-level reformulation. In *Proceedings of the IEEE/CVF Conference on Computer Vision and Pattern Recognition*, pages 11993–12002, 2020.

73. Shoukang Hu, Sirui Xie, Hehui Zheng, Chunxiao Liu, Jianping Shi, Xunying Liu, and Dahua Lin. Dsnas: Direct neural architecture search without parameter retraining. In *Proceedings of the IEEE/CVF Conference on Computer Vision and Pattern Recognition*, pages 12084–12092, 2020.

74. Roxana Istrate, Florian Scheidegger, Giovanni Mariani, Dimitrios S. Nikolopoulos, Costas Bekas, and A. Cristiano I. Malossi. TAPAS: train-less accuracy predictor for architecture search. *CoRR*, abs/1806.00250, 2018.

75. Jing Jiang, Fei Han, Qinghua Ling, Jie Wang, Tiange Li, and Henry Han. Efficient network architecture search via multiobjective particle swarm optimization based on decomposition. *Neural Networks*, 123:305–316, 2020.

76. Kirthevasan Kandasamy, Willie Neiswanger, Jeff Schneider, Barnabas Poczos, and Eric Xing. Neural architecture search with bayesian optimisation and optimal transport. *arXiv preprint arXiv:1802.07191*, 2018.

77. Aaron Klein, Stefan Falkner, Jost Tobias Springenberg, and Frank Hutter. Learning curve prediction with bayesian neural networks. In *5th International Conference on Learning Representations, ICLR 2017, Toulon, France, April 24–26, 2017, Conference Track Proceedings*. OpenReview.net, 2017.

78. Oliver Kramer. Evolution of convolutional highway networks. In *International Conference on the Applications of Evolutionary Computation*, pages 395–404. Springer, 2018.

79. Guohao Li, Guocheng Qian, Itzel C Delgadillo, Matthias Muller, Ali Thabet, and Bernard Ghanem. Sgas: Sequential greedy architecture search. In *Proceedings of the IEEE/CVF Conference on Computer Vision and Pattern Recognition*, pages 1620–1630, 2020.

80. Xiang Li, Chen Lin, Chuming Li, Ming Sun, Wei Wu, Junjie Yan, and Wanli Ouyang. Improving one-shot nas by suppressing the posterior fading. In *Proceedings of the IEEE/CVF Conference on Computer Vision and Pattern Recognition*, pages 13836–13845, 2020.

81. Chenxi Liu, Barret Zoph, Maxim Neumann, Jonathon Shlens, Wei Hua, Li-Jia Li, Li Fei-Fei, Alan Yuille, Jonathan Huang, and Kevin Murphy. Progressive neural architecture search. In *Proceedings of the European conference on computer vision (ECCV)*, pages 19–34, 2018.

82. Chenxi Liu, Barret Zoph, Jonathon Shlens, Wei Hua, Li-Jia Li, Li Fei-Fei, Alan L. Yuille, Jonathan Huang, and Kevin Murphy. Progressive neural architecture search. *CoRR*, abs/1712.00559, 2017.

83. Hanxiao Liu, Karen Simonyan, Oriol Vinyals, Chrisantha Fernando, and Koray Kavukcuoglu. Hierarchical representations for efficient architecture search. In *Proceedings of the 6th International Conference on Learning Representations*, 2018.

84. Hanxiao Liu, Karen Simonyan, and Yiming Yang. Darts: Differentiable architecture search. In *Proceedings of the 7th International Conference on Learning Representations*, 2019.

85. Yuqiao Liu, Yanan Sun, Bing Xue, Mengjie Zhang, and Gary G. Yen. A survey on evolutionary neural architecture search. *CoRR*, abs/2008.10937, 2020.

86. Pablo Ribalta Lorenzo and Jakub Nalepa. Memetic evolution of deep neural networks. In *Proceedings of the Genetic and Evolutionary Computation Conference*, pages 505–512, 2018.

87. Zhichao Lu, Ian Whalen, Vishnu Boddeti, Yashesh Dhebar, Kalyanmoy Deb, Erik Goodman, and Wolfgang Banzhaf. Nsga-net: neural architecture search using multi-objective genetic algorithm. In *Proceedings of the Genetic and Evolutionary Computation Conference*, pages 419–427, 2019.

88. Zhichao Lu, Ian Whalen, Yashesh Dhebar, Kalyanmoy Deb, Erik Goodman, Wolfgang Banzhaf, and Vishnu Naresh Boddeti. Multi-criterion evolutionary design of deep convolutional neural networks. *arXiv preprint arXiv:1912.01369*, 2019.

89. Renqian Luo, Fei Tian, Tao Qin, Enhong Chen, and Tie-Yan Liu. Neural architecture optimization. In *Advances in neural information processing systems*, pages 7816–7827, 2018.

90. Krzysztof Maziarz, Andrey Khorlin, Quentin de Laroussilhe, and Andrea Gesmundo. Evolutionary-neural hybrid agents for architecture search. *CoRR*, abs/1811.09828, 2018.

91. Andrew Y. Ng, Daishi Harada, and Stuart J. Russell. Policy invariance under reward transformations: Theory and application to reward shaping. In Ivan Bratko and Saso Dzeroski, editors, *Proceedings of the Sixteenth International Conference on Machine Learning (ICML 1999), Bled, Slovenia, June 27–30, 1999*, pages 278–287. Morgan Kaufmann, 1999.

92. Houwen Peng, Hao Du, Hongyuan Yu, Qi Li, Jing Liao, and Jianlong Fu. Cream of the crop: Distilling prioritized paths for one-shot neural architecture search, 2020.

93. Hieu Pham, Melody Y Guan, Barret Zoph, Quoc V Le, and Jeff Dean. Efficient neural architecture search via parameter sharing. In *Proceedings of the 35th International Conference on Machine Learning*, 2018.

94. Ning Qiang, Bao Ge, Qinglin Dong, Fangfei Ge, and Tianming Liu. Neural architecture search for optimizing deep belief network models of fmri data. In *International Workshop on Multiscale Multimodal Medical Imaging*, pages 26–34. Springer, 2019.

95. Esteban Real, Alok Aggarwal, Yanping Huang, and Quoc V Le. Regularized evolution for image classifier architecture search. In *Proceedings of the aaai conference on artificial intelligence*, volume 33, pages 4780–4789, 2019.

96. Esteban Real, Sherry Moore, Andrew Selle, Saurabh Saxena, Yutaka Leon Suematsu, Jie Tan, Quoc V Le, and Alexey Kurakin. Large-scale evolution of image classifiers. In *Proceedings of the 34th International Conference on Machine Learning-Volume 70*, pages 2902–2911. JMLR. org, 2017.

97. Martin Schrimpf, Stephen Merity, James Bradbury, and Richard Socher. A flexible approach to automated RNN architecture generation. *CoRR*, abs/1712.07316, 2017.

98. John Schulman, Filip Wolski, Prafulla Dhariwal, Alec Radford, and Oleg Klimov. Proximal policy optimization algorithms. *CoRR*, abs/1707.06347, 2017.

99. Han Shi, Renjie Pi, Hang Xu, Zhenguo Li, James Kwok, and Tong Zhang. Bridging the gap between sample-based and one-shot neural architecture search with bonas. *Advances in Neural Information Processing Systems*, 33, 2020.

100. Yao Shu, Wei Wang, and Shaofeng Cai. Understanding architectures learnt by cell-based neural architecture search. In *8th International Conference on Learning Representations, ICLR 2020, Addis Ababa, Ethiopia, April 26–30, 2020*, 2020.

101. Yanan Sun, Bing Xue, Mengjie Zhang, and Gary G Yen. Evolving deep convolutional neural networks for image classification. *IEEE Transactions on Evolutionary Computation*, 24(2):394–407, 2020.

102. Yehui Tang, Yunhe Wang, Yixing Xu, Hanting Chen, Chunjing Xu, Boxin Shi, Chao Xu, Qi Tian, and Chang Xu. A semi-supervised assessor of neural architectures, 2020.

103. Haiman Tian, Shu-Ching Chen, Mei-Ling Shyu, and Stuart Rubin. Automated neural network construction with similarity sensitive evolutionary algorithms. In *2019 IEEE 20th International Conference on Information Reuse and Integration for Data Science (IRI)*, pages 283–290. IEEE, 2019.

104. Łukasz Dudziak, Thomas Chau, Mohamed S. Abdelfattah, Royson Lee, Hyeji Kim, and Nicholas D. Lane. Brp-nas: Prediction-based nas using gcns, 2021.

105. Bin Wang, Yanan Sun, Bing Xue, and Mengjie Zhang. Evolving deep neural networks by multi-objective particle swarm optimization for image classification. In *Proceedings of the Genetic and Evolutionary Computation Conference*, pages 490–498, 2019.

106. Tao Wei, Changhu Wang, Yong Rui, and Chang Wen Chen. Network morphism. In *International Conference on Machine Learning*, pages 564–572, 2016.

107. Martin Wistuba, Ambrish Rawat, and Tejaswini Pedapati. A survey on neural architecture search. *arXiv preprint arXiv:1905.01392*, 2019.

108. Lingxi Xie and Alan Yuille. Genetic cnn. In *Proceedings of the IEEE international conference on computer vision*, pages 1379–1388, 2017.

109. Saining Xie, Alexander Kirillov, Ross B. Girshick, and Kaiming He. Exploring randomly wired neural networks for image recognition. In *2019 IEEE/CVF International Conference on Computer Vision, ICCV 2019, Seoul, Korea (South), October 27 - November 2, 2019*, pages 1284–1293. IEEE, 2019.

110. Sirui Xie, Hehui Zheng, Chunxiao Liu, and Liang Lin. Snas: stochastic neural architecture search. In *Proceedings of the 7th International Conference on Learning Representations*, 2019.

111. Yuhui Xu, Lingxi Xie, Xiaopeng Zhang, Xin Chen, Guo-Jun Qi, Qi Tian, and Hongkai Xiong. Pc-darts: Partial channel connections for memory-efficient architecture search. *arXiv preprint arXiv:1907.05737*, 2019.

112. Shen Yan, Yu Zheng, Wei Ao, Xiao Zeng, and Mi Zhang. Does unsupervised architecture representation learning help neural architecture search?, 2020.

113. Yibo Yang, Hongyang Li, Shan You, Fei Wang, Chen Qian, and Zhouchen Lin. ISTA-NAS: efficient and consistent neural architecture search by sparse coding. In *Advances in Neural Information Processing Systems 33: Annual Conference on Neural Information Processing Systems 2020, NeurIPS 2020, December 6–12, 2020, virtual*, 2020.

114. Zhaohui Yang, Yunhe Wang, Xinghao Chen, Boxin Shi, Chao Xu, Chunjing Xu, Qi Tian, and Chang Xu. Cars: Continuous evolution for efficient neural architecture search. *arXiv preprint arXiv:1909.04977*, 2019.

115. Shan You, Tao Huang, Mingmin Yang, Fei Wang, Chen Qian, and Changshui Zhang. Greedynas: Towards fast one-shot nas with greedy supernet, 2020.

116. Kaicheng Yu, Christian Sciuto, Martin Jaggi, Claudiu Musat, and Mathieu Salzmann. Evaluating the search phase of neural architecture search. In *8th International Conference on Learning Representations, ICLR 2020, Addis Ababa, Ethiopia, April 26–30, 2020*. OpenReview.net, 2020.

117. Arber Zela, Aaron Klein, Stefan Falkner, and Frank Hutter. Towards automated deep learning: Efficient joint neural architecture and hyperparameter search. *CoRR*, abs/1807.06906, 2018.

118. Miao Zhang, Huiqi Li, Shirui Pan, Xiaojun Chang, Zongyuan Ge, and Steven W Su. Differentiable neural architecture search in equivalent space with exploration enhancement. *Advances in Neural Information Processing Systems*, 33:6–12, 2020.

119. Miao Zhang, Huiqi Li, Shirui Pan, Xiaojun Chang, and Steven Su. Overcoming multi-model forgetting in one-shot nas with diversity maximization. In *Proceedings of the IEEE/CVF Conference on Computer Vision and Pattern Recognition*, pages 7809–7818, 2020.

120. Xiawu Zheng, Rongrong Ji, Qiang Wang, Qixiang Ye, Zhenguo Li, Yonghong Tian, and Qi Tian. Rethinking performance estimation in neural architecture search, 2020.

121. Zhao Zhong, Junjie Yan, Wei Wu, Jing Shao, and Cheng-Lin Liu. Practical block-wise neural network architecture generation. In *Proceedings of the IEEE conference on computer vision and pattern recognition*, pages 2423–2432, 2018.

122. Dongzhan Zhou, Xinchi Zhou, Wenwei Zhang, Chen Change Loy, Shuai Yi, Xuesen Zhang, and Wanli Ouyang. Econas: Finding proxies for economical neural architecture search, 2020.

123. Barret Zoph and Quoc V Le. Neural architecture search with reinforcement learning. In *Proceedings of the 5th International Conference on Learning Representations*, 2017.

124. Barret Zoph, Vijay Vasudevan, Jonathon Shlens, and Quoc V. Le. Learning transferable architectures for scalable image recognition. In *2018 IEEE Conference on Computer Vision and Pattern Recognition, CVPR 2018, Salt Lake City, UT, USA, June 18–22, 2018*, pages 8697–8710. IEEE Computer Society, 2018.

125. Ferran Alet, Martin F Schneider, Tomas Lozano-Perez, and Leslie Pack Kaelbling. Meta-learning curiosity algorithms. In *International Conference on Learning Representations*, 2019.

126. Hao-Tien Lewis Chiang, Aleksandra Faust, Marek Fiser, and Anthony Francis. Learning navigation behaviors end-to-end with autorl. *IEEE Robotics and Automation Letters*, 4(2):2007–2014, 2019.

127. Aleksandra Faust, Anthony Francis, and Dar Mehta. Evolving rewards to automate reinforcement learning. *arXiv preprint arXiv:1905.07628*, 2019.

128. Jörg KH Franke, Gregor Köhler, Noor Awad, and Frank Hutter. Neural architecture evolution in deep reinforcement learning for continuous control. *arXiv preprint arXiv:1910.12824*, 2019.

129. Peter Henderson, Riashat Islam, Philip Bachman, Joelle Pineau, Doina Precup, and David Meger. Deep reinforcement learning that matters. *arXiv preprint arXiv:1709.06560*, 2017.

130. Leslie Pack Kaelbling. The foundation of efficient robot learning. *Science*, 369(6506):915–916, 2020.

131. Junhyuk Oh, Matteo Hessel, Wojciech M Czarnecki, Zhongwen Xu, Hado van Hasselt, Satinder Singh, and David Silver. Discovering reinforcement learning algorithms. *arXiv preprint arXiv:2007.08794*, 2020.

132. Supratik Paul, Vitaly Kurin, and Shimon Whiteson.   Fast efficient hyperparameter tuning for policy gradient methods. In *Advances in Neural Information Processing Systems*, pages 4616–4626, 2019.

133. Frederic Runge, Danny Stoll, Stefan Falkner, and Frank Hutter. Learning to design rna. In *International Conference on Learning Representations*, 2018.

134. Jürgen Schmidhuber. Formal theory of creativity, fun, and intrinsic motivation (1990–2010). *IEEE Transactions on Autonomous Mental Development*, 2(3):230–247, 2010.

135. Richard S Sutton and Andrew G Barto. *Reinforcement learning: An introduction.* MIT press, 2018.

136. Yunhao Tang and Krzysztof Choromanski.   Online hyper-parameter tuning in off-policy learning via evolutionary strategies. *arXiv preprint arXiv:2006.07554*, 2020.

137. Zhongwen Xu, Hado P van Hasselt, and David Silver. Meta-gradient reinforcement learning. In *Advances in neural information processing systems*, pages 2396–2407, 2018.

138. Tom Zahavy, Zhongwen Xu, Vivek Veeriah, Matteo Hessel, Junhyuk Oh, Hado van Hasselt, David Silver, and Satinder Singh. Self-tuning deep reinforcement learning. *arXiv preprint arXiv:2002.12928*, 2020.

139. Cai, H., Zhu, L., Han, S.: Proxylessnas: Direct neural architecture search on target task and hardware. arXiv preprint arXiv:1812.00332 (2018)

140. Chau, T., Dudziak, Ł., Abdelfattah, M.S., Lee, R., Kim, H., Lane, N.D.: Brp-nas: Prediction-based nas using gcns. arXiv preprint arXiv:2007.08668 (2020)

141. Dong, J.D., Cheng, A.C., Juan, D.C., Wei, W., Sun, M.: Dpp-net: Device-aware progressive search for pareto-optimal neural architectures. In: Proceedings of the European Conference on Computer Vision (ECCV), pp. 517–531 (2018)

142. Jiang, Y., Wang, X., Zhu, W.: Hardware-aware transformable architecture search with efficient search space. In: 2020 IEEE International Conference on Multimedia and Expo (ICME), pp. 1–6. IEEE (2020)

143. Stamoulis, D., Cai, E., Juan, D.C., Marculescu, D.: Hyperpower: Power-and memory-constrained hyper-parameter optimization for neural networks. In: 2018 Design, Automation & Test in Europe Conference & Exhibition (DATE), pp. 19–24. IEEE (2018)

144. Tan, M., Chen, B., Pang, R., Vasudevan, V., Sandler, M., Howard, A., Le, Q.V.: Mnasnet: Platform-aware neural architecture search for mobile.   In: Proceedings of the IEEE/CVF Conference on Computer Vision and Pattern Recognition, pp. 2820–2828 (2019)

145. todo: Hw-nas-bench: Hardware-aware neural architecture search benchmark. In: todo, p. todo (2021)

146. Wang, H., Wu, Z., Liu, Z., Cai, H., Zhu, L., Gan, C., Han, S.: Hat: Hardware-aware transformers for efficient natural language processing.   arXiv preprint arXiv:2005.14187 (2020)

147. Wu, B., Dai, X., Zhang, P., Wang, Y., Sun, F., Wu, Y., Tian, Y., Vajda, P., Jia, Y., Keutzer, K.: Fbnet: Hardware-aware efficient convnet design via differentiable neural architecture search. In: Proceedings of the IEEE/CVF Conference on Computer Vision and Pattern Recognition, pp. 10734–10742 (2019)

148. Zhang, L., Yang, Y., Jiang, Y., Zhu, W., Liu, Y.: Fast hardware-aware neural architecture search.   In: Proceedings of the IEEE/CVF Conference on Computer Vision and Pattern Recognition Workshops, pp. 692–693 (2020)

149. Dong, X., Yang, Y.: Network pruning via transformable architecture search. arXiv preprint arXiv:1905.09717 (2019)

150. Aaron Clauset, Mark EJ Newman, and Cristopher Moore. Finding community structure in very large networks. *Physical review E*, 70(6):066111, 2004.

151. Peng Cui, Xiao Wang, Jian Pei, and Wenwu Zhu. A survey on network embedding. *IEEE Transactions on Knowledge and Data Engineering*, 31(5):833–852, 2018.

152. Anton Tsitsulin, Davide Mottin, Panagiotis Karras, Alexander Bronstein, and Emmanuel Müller. Netlsd: hearing the shape of a graph. In *Proceedings of the 24th ACM SIGKDD International Conference on Knowledge Discovery & Data Mining*, pages 2347–2356, 2018.

153. Ke Tu, Jianxin Ma, Peng Cui, Jian Pei, and Wenwu Zhu. Autone: Hyperparameter optimization for massive network embedding. In *Proceedings of the 25th ACM SIGKDD International Conference on Knowledge Discovery & Data Mining*, pages 216–225, 2019.
154. Ziwei Zhang, Peng Cui, and Wenwu Zhu. Deep learning on graphs: A survey. *IEEE Transactions on Knowledge and Data Engineering*, 2020.
155. Y. Gao, H. Yang, P. Zhang, C. Zhou, and Y. Hu. Graphnas: Graph neural architecture search with reinforcement learning. *arXiv preprint arXiv:1904.09981*, 2019.
156. X. Kou, B. Luo, H. Hu, and Y. Zhang. Nase: Learning knowledge graph embedding for link prediction via neural architecture search. *arXiv preprint arXiv:2008.07723*, 2020.
157. G. Li, G. Qian, I. C. Delgadillo, M. Muller, A. Thabet, and B. Ghanem. Sgas: Sequential greedy architecture search. In *Proceedings of the IEEE/CVF Conference on Computer Vision and Pattern Recognition*, pages 1620–1630, 2020.
158. M. Nunes and G. L. Pappa. Neural architecture search in graph neural networks. *arXiv preprint arXiv:2008.00077*, 2020.
159. W. Peng, X. Hong, H. Chen, and G. Zhao. Learning graph convolutional network for skeleton-based human action recognition by neural searching. In *AAAI*, pages 2669–2676, 2020.
160. Y. Zhao, D. Wang, X. Gao, R. Mullins, P. Lio, and M. Jamnik. Probabilistic dual network architecture search on graphs. *arXiv preprint arXiv:2003.09676*, 2020.
161. K. Zhou, Q. Song, X. Huang, and X. Hu. Auto-gnn: Neural architecture search of graph neural networks. *arXiv preprint arXiv:1909.03184*, 2019.

Chapter 2
Meta-Learning

Last decade has witnessed a prosperous development for supervised learning, i.e., learning tasks with given labels for model training. Supervised learning usually depends on large labeled datasets and trains a huge model with a large number of parameters from scratch. Thus, the requirement for data and computing resources is relatively high. However, there are many applications where data is difficult or expensive to collect, or computing resources are limited. Since the lack of training data, supervised learning is not suitable for these tasks and shows bad performances.

On the other hand, human can learn new concepts and skills quite fast and efficiently from few data. We human reference and reuse the methods that worked well in related tasks in the past, rather than directly start from scratch [1]. Kids who have seen dogs and fish only a few times can quickly tell them apart and people knowing how to ride a bike with high probability are able to figure out riding a motorcycle fast with little or even no demonstration. As such, by using the previous experience, we can quickly adapt to new tasks after several trials, which is known as the ability of " learning to learn".

For the sake of towards human-like learning, meta-learning which targets at simulating the concept of "learning to learn", provides a paradigm where machine learning models are built based on experience with related tasks. Meta-learning has been becoming a very hot research topic in both academy and industry since the year of 2017, covering many research communities including machine learning, computer vision, natural language processing, data mining and multimedia.

In this chapter, we provide an overview of meta-learning approaches, including both basic concepts and advanced techniques. We summarize meta-learning as a series of techniques that can learn prior experience across tasks in a systematic, data-driven manner. We define the problem of meta-learning in two views, i.e., *task distribution view* and *learner and meta-learner view*.

Task Distribution View A good meta-learning method should help the model f_θ gain the ability of learning to learn across tasks and improve its performance on a distribution of tasks $p(\tau)$, including potentially unseen tasks. The optimal model

© The Author(s), under exclusive license to Springer Nature Switzerland AG 2021
W. Zhu, X. Wang, *Automated Machine Learning and Meta-Learning for Multimedia*,
https://doi.org/10.1007/978-3-030-88132-0_2

parameters are:

$$\theta^* = \arg\min_{\theta} \mathbb{E}_{\tau_i \sim p(\tau)}[\mathcal{L}_{\theta}(\tau_i)]. \tag{2.1}$$

To implement the optimization goal Eq. (2.1), we usually sample M meta-train tasks from $p(\tau)$, with which we learn meta-knowledge ω. The meta-knowledge ω guides the optimization of the model across tasks, and can have different meanings, such as parameter initialization, gradients, and optimization strategy. The training and validation sets of a meta-train task are often called support set and query set. Formally, we denote the set used in the meta-train stage as $\mathcal{D}_{meta-train} = \left\{ (\mathcal{D}_{meta-train}^S, \mathcal{D}_{meta-train}^Q)^{(i)} \right\}_{i=1}^{M}$. After meta-train stage, the model gets the optimal meta-knowledge ω^*:

$$\omega^* = \arg\max_{\omega} \log p(\omega | \mathcal{D}_{meta-train}). \tag{2.2}$$

Similarly, we denote the N sampled tasks used in the meta-test stage as $\mathcal{D}_{meta-test} = \left\{ (\mathcal{D}_{meta-test}^S, \mathcal{D}_{meta-test}^Q)^{(i)} \right\}_{i=1}^{N}$. We use the learned meta-knowledge to train the model on each previously unseen task and get the optimal model parameters:

$$\theta^* = \arg\max_{\theta} \log p(\theta | \omega^*, \mathcal{D}_{meta-test}^S). \tag{2.3}$$

We can evaluate the meta-learning algorithm by the performance of $\theta^*(i)$ on the corresponding meta-test query set $\mathcal{D}_{meta-test}^Q$.

Learner and Meta-Learner View   Another common view is that meta-learning decomposes the process of parameter update into two stages: base-learning stage and meta-learning stage. As aforementioned, the dataset for each task $\tau_i \sim p(\tau)$ is divided into support set $S^{(i)}$ and query set $Q^{(i)}$. During base learning stage, an inner learner model f_θ is trained on the support set $S^{(i)}$ for solving a given specific task. During meta-learning stage, an outer meta-learner g_ω is applied to improve an outer objective $\mathcal{L}^{meta}$ that is calculated on the query set $Q^{(i)}$. The outer optimization problem contains the inner optimization as a constraint. Using this notation, the parameter ω of meta-learner g_ω can be regarded as meta-knowledge. The meta-learning algorithm can be formulated as follows:

$$\omega^* = \arg\min_{\omega} \sum_{i=1}^{N} \mathcal{L}^{meta}\left(\theta^{*(i)}(\omega), \omega, Q^{(i)}\right), \tag{2.4}$$

$$s.t. \quad \theta^{*(i)}(\omega) = \arg\min_{\theta} \mathcal{L}^{base}\left(\theta, \omega, S^{(i)}\right), \tag{2.5}$$

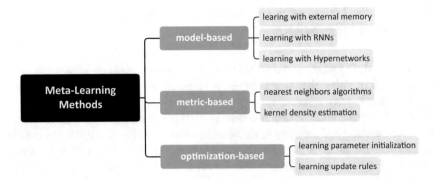

Fig. 2.1 A taxonomy of meta-learning methods

where $\mathcal{L}_{meta}$ and $\mathcal{L}_{base}$ refer to the outer and inner objective losses respectively, such as cross entropy in the case of few-shot classification.

In summary, this chapter presents a comprehensive overview of meta-learning approaches from three categories, i.e., model-based, metric-based, and optimization-based methods, as is illustrated in Fig. 2.1. We also discuss some recent advances on meta-learning, including dynamic meta-learning, multimodal meta-learning, meta reinforcement learning and meta-learning for graph.

2.1 Model-Based Methods

Model-based meta-learning models make no assumptions on task characteristics and configurations. Instead, it depends on a model specifically designed for fast learning—a model that can quickly update its parameters within just a few training steps. From the learner and meta-learner view, the rapid parameter update process can be considered as the inner optimization shown in Eq. (2.5). The model embeds the support set $S^{(i)}$ into its activation state, with predictions for data from the query set $Q^{(i)}$ being made based on this state. The outer prediction can be achieved by its internal architecture or controlled by another meta-learner model.

Some typical architectures including recurrent networks [2] and hypernetworks [3, 4], or external memory storage [5, 6] are usually applied in the outer-level stage to facilitate the learning process of models.

2.1.1   Meta-Learning with Memory-Augmented Neural Networks (MANN)

Santoro et al. [5] propose the MANN model, a neural network with augmented memory capabilities that can memorize information about previous tasks and leverage that to learn a learner l_{new}.

MANN takes Neural Turing Machine (NTM) [7] as a base model, and modifies the training setup and memory retrieval mechanisms. NTM consists of a controller and an external memory module. Interaction between both is through several read and write heads, which are responsible for retrieving representations from memory or placing them into memory, respectively. The read and write heads, actually work through soft attention mechanisms.

A. Task Setup

In order to use MANN for meta-learning tasks, we need to train it using a general learning strategy that the memory can encode and capture new task representations in a fast and stable way. In each training episode, the input at time step t consists of the truth label at the previous time step $t - 1$ and the input data $\mathbf{x}_t$, as the following form: $(\mathbf{x}_t, y_{t-1})$. With the one-step offset, the memory has to hold the current input until the label is present later and then retrieve the old information to solve new tasks. Under this intuition, MANN can memorize the information of a new dataset and adapt to a new task.

B. Read from Memory

Read attention is constructed purely based on the content similarity. We denote a key feature vector produced by the controller at the time step t as $\mathbf{k}_t$, which is either stored in a row of a memory matrix $\mathbf{M}_t$, or used to retrieve a particular memory, i, from a row; i.e., $\mathbf{M}_t(i)$. When retrieving a memory, a read-weight vector $\mathbf{w}_r^t$ of N elements is computed as the cosine similarity between the key vector $\mathbf{k}_t$ and every memory vector row $\mathbf{M}_t(i)$, normalized by softmax:

$$\mathbf{w}_t^r(i) = \text{softmax}(\frac{\mathbf{k}_t \cdot \mathbf{M}_t(i)}{\|\mathbf{k}_t\| \cdot \|\mathbf{M}_t(i)\|}). \tag{2.6}$$

The read vector, $\mathbf{r}_t$, is retrieved as a sum of memory records weighted by this weight vector:

$$\mathbf{r}_t = \sum_{i=1}^{N} \mathbf{w}_t^r(i)\mathbf{M}_t(i). \tag{2.7}$$

C. Write to Memory

A pure content-based memory writer called the LRUA module is utilized to make the model better encode relevant (i.e., recent) information and retrieve the memory. This module operates a lot like the cache replacement policy LRUA and makes MANN more suitable for the scenario of meta-learning. An LRUA write head prefers to write new memories to either the least used memory location or the most recently used memory location.

At time-stamp t, the usage weights $\mathbf{w}_t^u$ is a sum of current read and write vectors, in addition to the decayed last usage weight:

$$\mathbf{w}_t^u = \gamma \mathbf{w}_{t-1}^u + \mathbf{w}_t^r + \mathbf{w}_t^w, \tag{2.8}$$

where γ is a decay factor and $\mathbf{w}_t^r$ is computed as in Eq. (2.6). Besides, the least-used weights, $\mathbf{w}_t^{lu}$, can be computed using $\mathbf{w}_t^u$ as:

$$\mathbf{w}_t^{lu} = \mathbf{1}_{w_t^u(i) \leq m(\mathbf{w}_t^u, n)}, \tag{2.9}$$

where $m(\mathbf{w}_t^u, n)$ is the n-th smallest element in vector $\mathbf{w}_t^u$. The write vector $\mathbf{w}_t^w$ is an interpolation between the previous read weights and previous least-used weights. The interpolation parameter is the sigmoid of a hyperparameter α. To obtain the write weights $\mathbf{w}_t^w$, a learnable sigmoid gate parameter is used to compute a convex combination of $\mathbf{w}_t^r$ and $\mathbf{w}_t^{lu}$:

$$\mathbf{w}_t^w = \sigma(\alpha) \mathbf{w}_{t-1}^r + (1 - \sigma(\alpha)) \mathbf{w}_{t-1}^{lu}. \tag{2.10}$$

Finally, after the least used memory location, indicated by $\mathbf{w}_{lu}$, is set to zero, every memory row is updated:

$$\mathbf{M}_t(i) = \mathbf{M}_{t-1}(i) + \mathbf{w}_t^w(i)\mathbf{k}_t, \forall i. \tag{2.11}$$

2.1.2 Extensions and Variants of MANN

Tran et al. [8] combine MANN and Matching Networks [11] and use the LRUA module to improve the full context embedding in Matching Networks. Inspired from MANN, Rae et al. [9] recently propose a new memory-augmented neural network architecture, the Neural Bloom Filter, to meta-learn one-shot approximate set membership problems of varying structure. Rae et al. find that the meta-learning process of sampling tasks from a common distribution matches very well to some applications where many Bloom Filters are instantiated over different subsets of common data distribution. Therefore, the memory-augmented neural architecture has the natural ability to meta-learn these tasks. They corporate the classical Bloom

Filter with different memory-augmented neural networks and discovered some more compressive solution structures.

2.2 Metric-Based Methods

The core idea in metric-based meta-learning is to transfer information from the most similar task to the new task, where the task similarity can be measured by a distance function or a kernel function. It is similar to nearest neighbors algorithms (i.e., k-NN classifier and k-means clustering) and kernel density estimation. The predicted probability over a set of known labels y is a weighted sum of labels of support set samples, as shown in Eq. (2.12). The weight is generated by a kernel function k_θ, measuring the similarity between two data samples.

$$\mathbf{P}_\theta(y|\mathbf{X}, S) = \sum_{(\mathbf{x}_i, y_i) \in S} k_\theta(\mathbf{x}, \mathbf{x}_i) y_i. \tag{2.12}$$

A good kernel is crucial and problem-dependent. Here the outer-level learning corresponds to finding a kernel function $\omega = k_\theta(\mathbf{x}_j, \mathbf{x}_i)$ that builds appropriate representations to model the relationship between input data in the task space and facilitate problem-solving. As before ω is also learned on support sets $S^{(i)}$, and used for query sets $Q^{(i)}$.

In chronological order, some works, such as siamese networks [10], matching networks [11], prototypical networks [12], relation networks [13], and graph neural networks [14], are proposed to learn embedding vectors of input data explicitly and use them to design proper kernel functions in different ways.

2.2.1 Matching Networks

Matching Networks [11] aims to solve a k-shot classification problem, which is to learn a mapping from a small support set $S = \{x_i, y_i\}_{i=1}^k$ to a classifier $c_S(\hat{x})$, given an example $(\hat{x}, \hat{y}) \in Q$. The mapping $S \rightarrow c_S(\hat{x})$ is defined to be $P(\hat{y}|\hat{x}, S)$, where P is parameterized by a neural network.

Similar to other metric-based models, the classifier output is defined as a sum of labels of support samples weighted by an attention kernel $a(\hat{x}, x_i)$:

$$c_S(\hat{x}) = P(\hat{y}|\hat{x}, S) = \sum_{i=1}^k a(\hat{x}, x_i) y_i. \tag{2.13}$$

A. The Attention Kernel

The attention kernel depends on two embedding functions, f and g, for encoding the test sample and the support set samples respectively. The attention weight between two data points is calculated by a softmax over the cosine distance between the two embedding vectors:

$$a(\hat{x}, x_i) = \frac{\exp(\text{cosine}(f(\hat{x}), g(x_i)))}{\sum_{j=1}^{k} \exp(\text{cosine}(f(\hat{x}), g(x_j)))}. \tag{2.14}$$

B. Full Context Embeddings

In the simple version, if the embedding function is a neural network with a single data sample x_i as input, we can potentially set $f = g$. However, just taking a single data point as input might not be enough to efficiently encode the entire feature space. Therefore, the Matching Network model further proposes to embed the elements of the set through a function which takes as input the full set S in addition to x_i, i.e., g becomes $g(x_i, S)$, so that the learned embedding can be adjusted based on the relationship with other support samples. In detail, a bidirectional LSTM is used to encode x_i in the context of the support set S, considered as a sequence.

Another embedding function $f(\hat{x}, S)$ can be fixed via an LSTM with read-attention over the whole set S, whose inputs are equal to $f'(\hat{x})$ (f' is an embedding function, e.g., a CNN). The training process of the LSTM is defined as following recurrence over "processing" steps k:

$$\hat{h}_k, c_k = \text{LSTM}(f'(\hat{x}), [h_{k-1}, r_{k-1}], ck - 1), \tag{2.15}$$

$$h_k = \hat{h}_k + f'(\hat{x}), \tag{2.16}$$

$$r_{k-1} = \sum_{i=1}^{|S|} a(h_{k-1}, g(x_i))g(x_i), \tag{2.17}$$

$$a(h_{k-1}, g(x_i)) = \frac{\exp(h_{k-1}^T)g(x_i)}{\sum_{j=1}^{|S|} \exp(h_{k-1}^T)g(x_j)}. \tag{2.18}$$

Eventually, $f(\hat{x}, S) = h_K$ if we do K steps of "read", where h_k is as described in Eq. (2.16).

2.2.2   Prototypical Networks

Prototypical Networks [12] use an embedding function $f_\phi : \mathbb{R}^D \to \mathbb{R}^M$ to encode each input $\mathbf{x}_i \in \mathbb{R}^D$ into a M-dimensional feature vector, $\mathbf{c}_k \in \mathbb{R}^M$, which is also called *prototype*. A prototype feature vector is defined for every class $c \in C$, as the mean vector of the embedded support points belonging to its class:

$$\mathbf{c}_k = \frac{1}{|S_k|} \sum_{(\mathbf{x}_i, y_i) \in S_k} f_\phi(\mathbf{x}_i). \tag{2.19}$$

The distribution over classes for a given query input point $\mathbf{x}$ is a softmax over distances to the prototypes in the embedding space:

$$P_\phi(y = k | \mathbf{x}) = \text{softmax}(-d(f_\phi(\mathbf{x}), \mathbf{c}_k)) = \frac{\exp(-d(f_\phi(\mathbf{x}), \mathbf{c}_k))}{\sum_{k'} \exp(-d(f_\phi(\mathbf{x}), \mathbf{c}'_k))}, \tag{2.20}$$

where $d : \mathbb{R}^M \times \mathbb{R}^M \to [0, +\infty)$ can be any distance function as long as d is differentiable. In the paper, the squared Euclidean distance is applied.

The loss function is the negative log-likelihood: $\mathcal{J}(\phi) = -\log P_\phi(y = k | \mathbf{x})$.

2.2.3   Extensions and Variants of Prototypical Networks

Few-shot learning has become essential for producing models that achieve good generalization performance using few training examples. Metric scaling and metric task conditioning are theoretically proved to play an important role in changing the nature of few-shot algorithm parameter updates. Under this background, Oreshkin et al. [15] propose a task-dependent adaptive metric method called Tadam. Prototypical Networks are re-implemented and corporates with a task embedding network (TEN) block to get task-conditional representations in a task-dependent metric space.

Ren et al. [16] augment Prototypical Networks with the ability to use unlabeled examples when producing prototypes. These models can make use of unlabeled examples to improve the predictions, performing like a semi-supervised algorithm. Another work [17] also extends Prototypical Networks to semi-supervised situations.

Li et al. [18] present a method that learns a global class representation different from the episodic one in Prototypical Networks. During the training process, both base and novel samples are used to jointly learn the representation, to ensure global consistency.

2.3 Optimization-Based Methods

Optimization-based meta-learning methods include those where the inner-level task Eq. (2.5) is solved as an optimization problem, and the outer-level process focuses on extracting meta-knowledge ω required to improve optimization performance. Moreover, the meta-knowledge ω can have different meanings, such as meta-gradient, parameter initialization, and parameters of an update rule.

The outer optimization process can be performed by gradient descent [19–23], differentiating through the updates to the base model. Some works [2, 24, 25] use a temporal model to learn the parameter update rule. Other techniques like evolution strategies [26] and reinforcement learning [27] can also be applied to optimize the model parameters.

2.3.1 The Model Agnostic Meta-Learning (MAML) Algorithm

This section discusses the most famous optimization-based method, Model Agnostic Meta-learning (MAML) [20]. MAML is a fairly general optimization algorithm, compatible with any model that learns through gradient descent.

A model is represented by a parameterized function f_θ with parameters θ. At each iteration, it selects a batch of prior tasks τ_i from $p(\tau)$ and the dataset $(\mathcal{D}_{train}^{(i)}, \mathcal{D}_{test}^{(i)})$ associated with each task. For each task, it evaluates the gradient $\nabla_\theta \mathcal{L}_{\tau_i}(f_\theta)$ with respect to K examples in $\mathcal{D}_{train}^{(i)}$. The updated parameter vector θ_i is computed using one or more gradient descent updates. After one gradient update, $\theta_i' = \theta - \alpha \nabla_\theta \mathcal{L}_{\tau_i}(f_\theta)$.

Meta-update uses a different set of data, $\mathcal{D}_{test}^{(i)}$. The model parameters are trained by optimizing for the performance of $f_{\theta_i'}$ and the meta-objective can be formulated as follows:

$$\min_\theta \sum_{\tau_i \sim p(\tau)} \mathcal{L}_{\tau_i}(f_{\theta_i'}) = \sum_{\tau_i \sim p(\tau)} \mathcal{L}_{\tau_i}(f_{\theta - \alpha \nabla_\theta \mathcal{L}_{\tau_i}(f_\theta)}). \tag{2.21}$$

The goal of MAML is to learn a model parameter initialization that generalizes better to similar tasks. It aims to optimize the model parameters such that one or a small number steps of gradient descent can produce maximally effective behavior on a new task. The meta-optimization process is performed over the model parameters θ via stochastic gradient descent (SGD) across tasks, such that θ are updated as follows:

$$\theta \leftarrow \theta - \beta \nabla_\theta \sum_{\tau_i \sim p(\tau)} \mathcal{L}_{\tau_i}(f_{\theta_i'}). \tag{2.22}$$

After each iteration, the model parameters θ are in the direction in which they would have been easier to update, such that it becomes a better start point to finetune any other similar tasks.

2.3.2 Extensions and Variants of MAML

The MAML meta-gradient update involves a gradient through a gradient. Computationally, this requires an additional backward pass to compute second-order derivatives and leads to the challenge of differentiating through a graph of potentially thousands of inner optimization steps. To make the computation less expensive, some extension works present different solutions, such as first-order approximations of MAML. REPTILE [28] works by repeatedly sampling a task and training on it by multiple gradient descent steps within each iteration. It executes stochastic gradient update for k steps on the loss of a given task $\mathcal{L}_{\tau_i}$ starting from initial parameter ϕ. The meta-optimization process is defined as $\phi \leftarrow \phi + \epsilon \frac{1}{n} \sum_{i=1}^{n} (\tilde{\phi}_i - \phi)$, instead of calculating second-order derivatives. Finally, the parameter vector eventually moves in the direction of the sum of weights obtained from each task.

Another key challenge with MAML is that the number of parameters to be solved in the outer optimization is as many as the number in the inner optimization. For large CNNs, it may be potentially up to hundreds of millions. Moreover, using the same parameter initialization for all tasks may not work well and lead to sub-optimal performance for each task. Therefore, a series of works appear on isolating a subset of parameters to meta-learn and taking some task-specific adaptation. Lee et al. [29] propose a method that enables the meta-learner to learn a subspace on each layer's activation space, on which the task-specific learner performs gradient descent. Additionally, a meta-learned distance metric is applied to warp the activation space to be more sensitive to task identity. LEO [30] utilizes an encoder-decoder architecture to learn a data-dependent latent generative representation and performs the gradient-based adaptation procedure in a low-dimensional space decoupled from the space of model parameters.

Some researchers also combine MAML with Bayesian learning. Grant et al. [31] interpret MAML as a probability inference method in a hierarchical Bayesian model. Finn et al. [32] propose a new derivation and extension of MAML, and interpret a task-specific gradient update as a posterior inference process under variational inference framework. Wang et al. [33] place meta-learning on the space of probability measures, and introduce a Bayesian meta sampling framework consisting of a meta sampler and a sample adapter.

Many other extensions and variants of MAML are proposed to solve the limited or noisy label problem [34, 35], or realize better generalization and fast adaption [36–38], etc.

2.4 Advances on Meta-Learning

In this section, we will introduce some advances on meta-learning, including dynamic meta-learning, multimodal meta-learning, meta reinforcement learning and meta-learning for graph. These topics are emerging directions in meta-learning research and we believe they may inspire the readers for future investigations.

2.4.1 Dynamic Meta-Learning

Meta-learning is an important and long-standing issue of Artificial Intelligence (AI) that focuses on enabling agents to efficiently learn new tasks with the mechanism of learning to learn. With accumulating meta knowledge [42], meta-learning models can build self-adaptive learners using algorithms that improve the performance on tasks [43]. Dynamic meta-learning is an approach that is concerned with learning the characteristics of time-series problems and often has dynamic environments or task-unsegmented settings for new or unseen scenarios, which usually has a little difference with the traditional meta-learning framework. The scope of dynamic meta-learning is narrower and more precise than meta-learning and focuses on some more challenging problems under-addressed in the meta-learning literature. Specifically, the methods of meta-learning generally focus on general multi-task problems, but the methods of dynamic meta-learning can continuously learn and adapt in some non-stationary environments, which are more common in the real world. Moreover, although meta-learning has been recently shown promising as an effective strategy for learning to learn within new tasks, it has focused on the task segmented setting as usual. It is still an important problem of dynamic meta-learning that how to enable meta-learning framework into task-unsegmented settings to operate directly on some tasks such as time series.

A. Dynamic Meta-Learning on Dynamical Environments

In the real world, the environments are often non-stationary and change dynamically driven by some underlying and unobserved physical dynamics. In traditional meta-learning literature, tasks on both training and testing are split separately, and the boundary is obvious. However, for dynamic meta-learning, it is expected that the agent can continuously learn and adapt to the changes in the dynamic environment at the testing time even without enough data between the changes [39]. This can be seen as a significant milestone on the path towards general intelligence.

A famous method, gradient-based model-agnostic meta-learning (MAML) [49] has been proposed and shown impressive performance in meta-learning tasks. Then this method is extended into dynamically changing tasks successfully [39].

For the classical meta-learning method, e.g., MAML, it aims to find a solution that can generate a good policy for solving the tasks. Under policy π_θ, it can get K trajectories $\tau_\theta^{1:K}$ from $T \sim D(T)$. Then the goal is to minimize the expected subsequent objective on task T with the policy π_ϕ. So the task-specific loss function L_T is as follows:

$$L_T(\tau_\theta^{1:K}) = \frac{1}{K} \sum_{k=1}^{K} L_T(\tau_\theta^k), \tag{2.23}$$

and

$$\phi = \theta - \alpha \nabla_\theta L_T(\tau_\theta^{1:K}), \tag{2.24}$$

where α is the step of adaptation update that is parametrized by θ. Finally, the objective is:

$$\min_\theta \mathbb{E}_{T \sim D(T)} \left[\mathcal{L}_T(\theta) \right], \tag{2.25}$$

where

$$\mathcal{L}_T(\theta) = \mathbb{E}_{\tau_\theta^{1:K} \sim p_T(\tau|\theta)} \left[\mathbb{E}_{\tau_\phi \sim p_T(\tau|\phi)} \left[L_T\left(\tau_\phi\right) \mid \tau_\theta^{1:K}, \theta \right] \right]. \tag{2.26}$$

Finally, the gradient of $\mathcal{L}$ is as follows:

$$\nabla_\theta \mathcal{L}_T(\theta) = \mathbb{E}_{\tau_\theta^{1:K} \sim P_T(\tau|\theta), \tau_\phi \sim P_T(\tau|\theta)} \left[L_T\left(\tau_\phi\right) \left[\nabla_\theta \log \pi_\phi\left(\tau_\phi\right) + \nabla_\theta \sum_{k=1}^{K} \log \pi_\theta\left(\tau_\theta^k\right) \right] \right]. \tag{2.27}$$

Different from the general meta-learning tasks that often make no assumptions about the distribution of tasks $D(T)$, this dynamic meta-learning method treats it as a series of continuous tasks on a certain time scale, namely the tasks refer to an underlying dynamics of the environment. So in this scenario, these tasks can be denoted by a Markov chain (MC), and the objective of this dynamic meta-learning method is as follows for minimizing the expected loss over the chain of tasks:

$$\min_\theta \mathbb{E}_{p(T_0), p(T_{i+1}|T_i)} \left[\sum_{i=1}^{L} \mathcal{L}_{T_i, T_{i+1}}(\theta) \right], \tag{2.28}$$

where L is the length of the Markov chain of tasks, $p(T_0)$ is the initial probabilities and $p(T_{i+1}|T_i)$ denotes the transition probabilities, and $\mathcal{L}_{T_i, T_{i+1}}(\theta)$ represents the objective of step T_i to T_{i+1} that depends on the meta-learning process.

For dynamic meta-learning, an important point is the adaptation updates which are optimal regarding the Markovian transitions between the tasks. So the dynamic meta-loss on a pair of consecutive tasks can be defined as:

$$\mathcal{L}_{T_i,T_{i+1}}(\theta) = \mathbb{E}_{\tau_{i,\theta}^{1:K} \sim p_{T_i}(\tau|\theta)} \left[\mathbb{E}_{\tau_{i+1,\phi} \sim p_{T_{i+1}}(\tau|\phi)} \left[L_{T_{i+1}} \left(\tau_{i+1,\phi} \right) \mid \tau_{i,\theta}^{1:K}, \theta \right] \right].$$
(2.29)

As described above, the main difference is from the $\mathcal{L}$ between classical and dynamic meta-learning. The $\tau_{i,\theta}^{1:K}$ is from the current task and is used to improve for the next task. To construct parameters of the policy for task T_{i+1}, meta-gradient steps with adaptive step sizes are as follows:

$$\phi_i^0 = \theta, \quad \tau_\theta^{1:K} \sim p_{T_i}(\tau \mid \theta)$$

$$\phi_i^m = \phi_i^{m-1} - \alpha_m \nabla_{\phi_i^{m-1}} L_{T_i} \left(\tau_{i,\phi_i^{m-1}}^{1:K} \right), \quad m = 1, \ldots, M-1$$
(2.30)

$$\phi_{i+1} = \phi_i^{M-1} - \alpha_M \nabla_{\phi_i^{M-1}} L_{T_i} \left(\tau_{i,\phi_i^{M-1}}^{1:K} \right),$$

where M is the number of steps. Finally, the gradient of $\mathcal{L}$ is:

$$\nabla_{\theta,\alpha} \mathcal{L}_{T_i,T_{i+1}}(\theta, \alpha) =$$
$$\mathbb{E}_{\substack{i,\phi \sim P_{T_i}(\tau \mid \theta) \\ \tau_{i+1,\phi} \sim P_{T_{i+1}}(\tau \mid \phi)}} \left[L_{T_{i+1}} \left(\tau_{i+1,\phi} \right) \left[\nabla_{\theta,\alpha} \log \pi_\phi \left(\tau_{i+1,\phi} \right) + \nabla_\theta \sum_{k=1}^{K} \log \pi_\theta \left(\tau_{i,\theta}^k \right) \right] \right].$$
(2.31)

This method is an important work of dynamic meta-learning using meta-learning in dynamic and non-stationary environments, which is suitable for continuous adaptation. Since the dynamics, a series of tasks are treated as a sequence and the method tries to model the dependencies between consecutive tasks to adapt to the dynamics when testing.

B. Dynamic Meta-Learning on Task-Unsegmented Settings

Recently, traditional meta-learning methods often have an offline meta-training stage, where a distribution of tasks is optimized for improving performance on new or unseen tasks. One of the common limitations is that they can only be utilized on tasks that are already segmented in advance. In some applications, task segmentation is expensive and even unavailable. However, dynamic meta-learning does not need this strong assumption and the agent can learn to learn even when tasks are unsegmented and the change of tasks is unseen. It considers that the agent should learn to adapt in a changing environment where the tasks can switch along with the environment and the switching process may not be directly detected [41]. So dynamic meta-learning aims to enable the dynamic power into meta-learning on some task-unsegmented settings. Instead of requiring the segmented datasets in the training stage, it can operate directly on time series where the task undergoes the unobserved change.

When we adopt meta-learning framework on these problems, the learning agent is fed with the input time series data x_t and needs to output the prediction probability $p(\hat{y}_t|x_t)$. Similar to the common literature of meta-learning, these data points are sampled from the distribution regarding the task T_t, i.e.,

$$p(x_t, y_t|T_t) = p(y_t|x_t, T_t)p(x_t|T_t). \tag{2.32}$$

Finally, the optimization problem is making the prediction $\hat{y}_t$ close to y_t, and the objective [41] is as follows:

$$\min_{\theta} \quad \mathbb{E}\left[\sum_{t=1}^{\infty} -\log p_{\theta}\left(y_t \mid x_{1:t}, y_{1:t-1}\right)\right],$$

$$\text{subject to } x_t, y_t \sim p(x_t, y_t|T_t), \quad T_t = \begin{cases} T_{t-1} & \text{w.p. } 1-\lambda, \\ T_{t,\text{new}} & \text{w.p. } , \lambda \end{cases} \tag{2.33}$$

$$T_1 \sim D(T), \quad T_{t,\text{new}} \sim D(T).$$

Here, $D(T)$ is the distribution of task and θ is the optimized parameter. Compared with the classical meta-learning framework, this method accesses the time-series data directly rather than the pre-segmented tasks, which is the key idea of dynamic meta-learning.

As one of the dynamic meta-learning methods, MOCA [41], adopting Bayesian changepoint detection, enables the dynamic meta-learning algorithms on this task. In detail, the recursive Bayesian filtering algorithm is used for run length in the conditional and joint density estimation setting. Then, the classical meta-learning method with parameters θ is utilized to provide an underlying predictive model conditioned on the run length. For the first stage, namely Bayesian run-length filtering, the method maintains a belief over possible run lengths r and uses b_t to denote the updated belief before observing data at that time step. At time t, the agent, fed with the input x_t, needs to output the prediction $p(y_t|x_{1:t}, y_{1:t-1})$, and then y_t is observed. The Bayesian update rule is:

$$\begin{aligned} b_t\left(r_t \mid x_t\right) &= p\left(r_t \mid x_{1:t}, y_{1:t-1}\right) = Z^{-1}p\left(r_t, x_t \mid x_{1:t-1}, y_{1:t-1}\right) \\ &= Z^{-1}p\left(x_t \mid x_{1:t-1}, y_{1:t-1}, r_t\right)p\left(r_t \mid x_{1:t-1}, y_{1:t-1}\right) \\ &= Z^{-1}p_{\theta}\left(x_t \mid \eta_{t-1}[r_t]\right)b_t\left(r_t\right), \end{aligned} \tag{2.34}$$

where Z is the normalization constant and $\eta[r]$ is the posterior statistics. After getting the y_t, the belief becomes:

$$b_t\left(r_t \mid x_t, y_t\right) = Z^{-1}p_{\theta}\left(y_t \mid x_t, \eta_{t-1}[r_t]\right)b_t\left(r_t \mid x_t\right). \tag{2.35}$$

Finally, according to the dynamic meta-learning assumptions, $b_{t+1}(r_{t+1})$ is derived by propagating this forward over time:

$$b_{t+1}\left(r_{t+1}\right) = p\left(r_{t+1} \mid \boldsymbol{x}_{1:t}, \boldsymbol{y}_{1:t}\right) = \sum_{r_t} p\left(r_{t+1}, r_t \mid \boldsymbol{x}_{1:t}, \boldsymbol{y}_{1:t}\right)$$

$$= \sum_{r_t} p\left(r_{t+1} \mid r_t, \boldsymbol{x}_{1:t}, \boldsymbol{y}_{1:t}\right) p\left(r_t \mid \boldsymbol{x}_{1:t}, \boldsymbol{y}_{1:t}\right) \tag{2.36}$$

$$= \sum_{r_t} p\left(r_{t+1} \mid r_t\right) b_t\left(r_t \mid \boldsymbol{x}_t, \boldsymbol{y}_t\right).$$

For the next stage, benefit from the Bayesian filtering algorithm, the method can back-propagate with the change point detection, and optimize the underlying predictive model directly. It is worth noting that the model can be any meta-learning model that admits a probabilistic interpretation. The prediction can be formulated as follows:

$$p_\theta\left(\hat{\boldsymbol{y}}_t \mid \boldsymbol{x}_{1:t}, \boldsymbol{y}_{1:t-1}\right) = \sum_{r_t=0}^{t-1} b_t\left(r_t \mid \boldsymbol{x}_t\right) p_\theta\left(\hat{\boldsymbol{y}}_t \mid \boldsymbol{x}_t, \boldsymbol{\eta}_{t-1}\left[r_t\right]\right). \tag{2.37}$$

After given the true label y_t, the negative log-likelihood loss can be computed. The belief over run length and the posterior statistics can be updated. A recursive update rule for the posterior statistics allows these parameters to be computed efficiently as follows:

$$\boldsymbol{\eta}_t[r] = h\left(\boldsymbol{x}_t, \boldsymbol{y}_t, \boldsymbol{\eta}_{t-1}[r-1]\right), \quad \forall r = 1, \ldots, t. \tag{2.38}$$

The algorithm finally processes the data iteratively from $t = 1$ to T and performs gradient descent update to parameter θ of dynamic meta-learning.

Summary Dynamic meta-learning broadens the ability of the base meta-learning methods to the problems of time-series data, namely the dynamic tasks or even unsegmented tasks. Under these settings, the tasks are not unrelated but evolve, so that they are more complicated and challenging but more suitable for the applications in real-world environments. It is a significant milestone on the path towards general intelligence in the field of Artificial Intelligence.

2.4.2 Multimodal Meta-Learning

Current meta-learning methods usually focus on a series of similar tasks with only a single signal (e.g., text, audio, or visual). However, there are different combinations of modalities for signals or tasks in real-world applications. Thus, some advanced works go one step ahead by proposing several multimodal meta-learning methods.

Modality-Level Multimodal Meta-Learning Multimodal inputs provide more information and could be used to enhance the meta-learning methods. For example, visual and text features could be complementary to each other—while different concepts (texts) may share similar visual features, different visual signals may also refer to the same concepts. Chen et al. propose an adaptive cross-modal meta-learning mechanism, which can adaptively combine information from both visual and textual modalities according to new image categories to be learned [46]. The model is built on top of metric-based meta-learning methods like prototypical networks [47] and incorporates language structure, containing label embeddings of all categories. They model the new cross-modal prototype representation as a convex combination of the two modalities. The experimental results show their model can effectively adjust its focus on the two modalities and outperform some uni-modality few-shot learning methods and modality-alignment methods, especially in the case when the number of shots is very small.

Task-Level Multimodal Meta-Learning Most of the current meta-learning methods focus on unimodal task distribution, which contains a series of similar tasks (e.g., classifying different combinations of digits). However, the task distribution can also be multimodal, which contains tasks from different input and label domains (e.g., classifying digits vs. classifying cats). Risto et al. propose a multimodal model-agnostic meta-learning (MMAML) via task-aware modulation [44]. MMAML builds upon Model-Agnostic Meta-Learning (MAML) algorithm [45] and consists of a modulation network and a task network. The modulation network predicts the task mode identity, which is used as an input by the task network. Then, the task network is further adapted to the task using gradient-based optimization. They conduct experiments on few-shot regression, image classification, and reinforcement learning tasks. Take image classification as an example, they use five different datasets as different modes and show the superiority of the proposed MMAML method.

2.4.3 Meta Reinforcement Learning

Similar to previous meta-learning tasks, Meta Reinforcement Learning also aims to develop an agent that can solve unseen tasks fast and efficiently.

Consider a distribution $\tau_i \sim P(\tau)$ over tasks, and each task τ_i formularized as a Markov Decision Processes (MDP) $M_i = (S, A, P_i, R_i)$, with shared state space S and action space A, and different transition probability P_i and reward function R_i. Denote π_{ϕ_i} to be policy parametered by ϕ_i, where $\phi_i = f_\theta(M_i)$ is the result learned by adaptation model f_θ from MDP M_i for task τ_i. To maximize the expected reward while dealing with a new tasks, the problem of meta RL can be viewed as a bilevel optimization problem as follows:

$$\max_\theta \sum_{\tau_i} \mathbb{E}_{\pi_{\phi_i}(\tau)}[R(\tau)] \quad s.t. \phi_i = f_\theta(M_i). \tag{2.39}$$

Algorithm 4 General meta-RL outline

---

while training **do**

    Adaptation loop (i.e., $f_\theta(M_i)$):

    Sample task τ_i from distribution $P(\tau)$, and collect data D_i from MDP M_i.

    Compute the adaptation loop $\phi_i = f(\theta, D_i)$.

    Collect data D'_i by policy π_{ϕ_i}.

    Meta training loop:

    Update θ according to loss of the adated policy $l_{\phi_i}(D'_i)$.

end while

To solve the above bilevel optimization problem, meta reinforcement learning has an outer loop (meta-learning loop) that learns the meta knowledge across different tasks, and an inner loop (adaptation loop) to adapt to individual tasks (as illustrated in Algorithm 4). Note that a meta reinforcement learning agent can run more than one round of adaptation loop, and can update across a batch of tasks in a meta training loop. Respectively, the objective of Eq. (2.39) represents the outer loop and the constraint is the inner loop.

Follow the general outline, since the outer loop is similar to vanilla reinforcement learning, the key point is how to define the inner loop specifically. There are two main types of solutions with the defined meta reinforcement learning problem: one implements the policy as a recurrent network and trains across a set of tasks; one aims to learn a parameter initialization from which fine-tuning for a new task.

Recurrence Solutions Recurrent networks are verified to support meta-learning in a fully supervised context by more previous works [50, 52]. Wang et al. [53] and Duan et al. [48] extend this approach to the RL setting by utilizing LTSM and RNN networks to construct the adaptation loop, and simultaneously propose Meta-RL and RL^2 algorithm. Compared to the standard RL that optimizes the policy $\pi : s_t \to \Delta(A)$, which maps a state in S to a distribution over action A, the policy in Meta-RL and RL^2 is a function that maps trajectory $\{s_0, a_0, s_1, a_1, \ldots, s_t\}$ to this distribution.

The procedures of agent-environment interactions are similar in Meta-RL and RL^2 algorithm. The framework of the interaction between a meta reinforcement learning agent and the environment is illustrated in Fig. 2.2. In each trial, a separate MDP is drawn from $P(\tau)$, and has multiple episodes. For each episode, the state is initialized as state s_0 drawn from the initial state distribution specific to the corresponding MDP. In each time t, after receiving an action a_t selected by policy π_{h_t}, the environment computes reward r_t, steps forward, and computes the next state s_{t+1}, until the episode is terminated. If so, it resets the state as s_0 and sets termination flag d_t to 1, but keeps iterating the hidden state h_t before the trial ends. Note that the next state s_{t+1}, action a_t, reward r_t, and termination flag d_t, are concatenated to form the input to the policy π_{h_t}, with which the agent generates the next hidden state h_{t+2} and action a_{t+1} under the condition on the hidden state h_{t+1}.

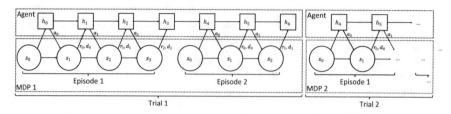

Fig. 2.2 The process of interaction between an agent and the environment of Meta-RL [53] and RL^2 [48]

Algorithm 5 Recurrence solution

---
while training **do**
    Adaptation loop (i.e., $f_\theta(M_i)$):
    Sample task τ_i from distribution $P(\tau)$,
    Initialize hidden state $h_0 = 0$.
    Sample (s_t, a_t, s_{t+1}, r_t) from π_{h_t} and let $D_i = D_i \cup \{(s_t, a_t, s_{t+1}, r_t)\}$.
    Update the hidden state by recurrence network:
        $h_{t+1} = f_\theta^{LSTM}(h_t, s_t, a_t, s_{t+1}, r_t)$ (Meta-RL).
        $h_{t+1} = f_\theta^{RNN}(h_t, s_t, a_t, s_{t+1}, r_t)$ (RL^2).

    Meta training loop:
    Update θ according to loss of the policy $l_{\pi_h}(D_i)$.
end while

The pseudocode of recurrence solution is illustrated in Algorithm 5, where the Meta-RL utilizes the LSTM to iterate the hidden state and the RL^2 uses RNN. Note that the adaptation loop can run multiple rounds, as previously defined. Since the underlying MDP changes across trials, different strategies are required for different MDPs. Then the recurrence creates a structure that encourages the agent to adapt itself across different MDPs, from which the adaptation to different tasks can be learned by the meta training loop. Hence, Algorithm 5 gives us a solution to learn a fast adaptation reinforcement learning algorithm.

Optimization Solutions Since a better initial that is closed to the optimal would accelerate the training, another idea is to find the initial model parameters that can be fast adapted to the new tasks. MAML [49] and REPTILE [28] methods train multiple steps for each task in the adaptation loop and then update the meta-model in the meta training loop.

As introduced previously, the key idea of the optimization solution is to run the adaptation loop of each sampled task and then optimize the adapted models from different tasks together in the meta training loop. The main difference is that optimization-based meta reinforcement learning needs to sample data from both the adapted policy and the meta policy, as illustrated in Algorithm 6, while the supervised meta-learning only needs to sample uniformly from one dataset. The same with supervised meta-learning, REPTILE can be viewed as an approximate method

---

Algorithm 6 Optimization solution

---

while training **do**
    Adaptation loop (i.e., $f_\theta(M_i)$):
    Sample task τ_i from distribution $P(\tau)$,
    Sample data D_i consists of k episodes from policy π_θ
    Update adapted parameters $\phi_i = \theta - \alpha \nabla_\theta l_{\pi_\theta}(D_i)$.
    Sample data D_i' consists of k episodes from policy π_{ϕ_i}

    Meta training loop:
    Update meta policy parameters:
        $\theta = \theta - \beta \nabla_\theta \sum_i l_{\pi_{\phi_i}}(D_i')$ (MAML)
        $\theta = \theta + \beta \sum_i (\phi_i - \theta)$ (Reptile)
end while

---

that avoids the second-order derivation in MAML to decrease the computational cost.

2.4.4 Meta-Learning for Graph

In Sect. 1.3.4, we have introduced how to conduct AutoML for graphs. In this section, we introduce the recent advances of meta-learning for graphs. Most tasks for graphs can be broadly divided into three categories: node-level, edge-level, and graph-level, depending on whether the tasks are associated with individual nodes, pairs of nodes (edges), or (sub)graphs, respectively [58].

Before detailing the meta-learning methods, we first introduce some backgrounds of graph neural networks (GNNs), which are generalizations of neural networks to graph data and have achieved promising performance in a wide range of graph tasks [62]. Consider a graph $\mathcal{G} = (\mathcal{V}, \mathcal{E}, \mathcal{F})$, where $\mathcal{V} = \{v_1, v_2, \ldots v_{|\mathcal{V}|}\}$ is a set of nodes, $\mathcal{E} \subseteq \mathcal{V} \times \mathcal{V}$ is a set of edges, and $\mathcal{F}$ are node features. The message-passing framework of GNNs is formulated as follows [57]:

$$\mathbf{m}_i^l = \mathrm{AGG}\left(\left\{\mathbf{h}_j^l, \forall v_j \in \mathcal{N}(i)\right\}\right), \tag{2.40}$$

$$\mathbf{h}_i^{l+1} = \mathrm{UPDATE}\left(\mathbf{m}_i^l, \mathbf{h}_i^l\right), \tag{2.41}$$

where $\mathbf{h}_i^l$ are the hidden representations of a node v_i in the lth layer, $\mathcal{N}(i) = \{v_j | (v_i, v_j) \in \mathcal{E}\}$ represents the neighborhoods of node v_i, $\mathbf{m}_i^l$ are messages by aggregating information from the neighborhoods, $\mathrm{AGG}(\cdot)$ is the aggregation function, and $\mathrm{UPDATE}(\cdot)$ is the updating function. In a nutshell, nodes exchange information with their immediate neighborhoods to update their representations. Node representations are usually initialized as the node features $\mathbf{h}_i^0 = \mathcal{F}(v_i)$.

Next, we introduce meta-learning for different categories of graph tasks. Semi-supervised node classification is a typical example of node-level tasks, where the goal is to predict the labels of nodes given a partially labeled node set. Meta-GNN [63] is proposed to combine meta-learning with GNNs to tackle the node classification problem in few-shot settings, where meta-learning can be utilized to transfer knowledge from meta-training classes. Specifically, Meta-GNN adopts the episodic paradigm to sample similar tasks as the few-shot setting in the meta-testing phase and uses MAML to meta-train the parameters of GNNs. In this way, the GNNs can learn to quickly adapt to new classes given a few samples.

Link prediction, aiming to predict which pairs of nodes are more likely to form edges, is a representative edge-level task. Meta-Graph [54] studies how to apply meta-learning to the link prediction problem. Specifically, the overall learning framework is also based on MAML of using second-order gradients to obtain good initializations of GNNs. To further bootstrap fast adaptation in cases where multiple graphs are adopted, i.e., the training, validation, and testing data can all contain various graphs without overlapping, Meta-Graph proposes to learn a graph signature function and adds the signature to the GNN model, i.e.,

$$\mathbf{m}_i^l = \text{AGG}\left(\left\{\mathbf{h}_j^l, \forall v_j \in \mathcal{N}(i)\right\}, \mathbf{s}_{\mathcal{G}^k}\right), \tag{2.42}$$

$$\mathbf{h}_i^{l+1} = \text{UPDATE}\left(\mathbf{m}_i^l, \mathbf{h}_i^l, \mathbf{s}_{\mathcal{G}^k}\right), \tag{2.43}$$

where $\mathbf{s}_{\mathcal{G}^k}$ is the signature of graph $\mathcal{G}^k = \left(\mathcal{V}^k, \mathcal{E}^k, \mathcal{F}^k\right)$ such that $v_i \in \mathcal{V}^k$. In this way, the model has extra flexibility in learning how to map the graph structure to an effective initialization point. To learn $\mathbf{s}_{\mathcal{G}^k}$, another GNN is applied (to disambiguate with the GNN in the link prediction model, we use $\mathbf{u}$ to denote node representations in this GNN) and the sum pooling is used to sum the node representations into the graph representations:

$$\mathbf{s}_{\mathcal{G}^k} = \text{MLP}(\sum\nolimits_{v_i \in \mathcal{V}^k} \mathbf{u}_i^L), \tag{2.44}$$

where $\mathbf{u}_i^L$ are the final node representations of node v_i in the GNN (assuming the GNN has L layers) and MLP is a multi-layer perceptron. All parameters are differentiable and can be learned in an end-to-end manner.

AS-MAML [60] proposes a meta-learning framework for graph classification by adopting the MAML paradigm as other methods. Instead of assuming that the inner adaptation steps in MAML are fixed for all the graphs, AS-MAML proposes an adaptive step controller to learn optimal adaptation steps to accommodate to the fact that graphs have arbitrary sizes and structures. Intuitively, if nodes can be well reconstructed by their neighborhoods, there is no need for further learning since messages are well dispersed among the nodes. Specifically, we can formulate this intuition by defining a metric named ANI (average node information):

$$\text{ANI} = \frac{1}{|\mathcal{V}|} \sum\nolimits_{v_i \in \mathcal{V}} \mathcal{D}\left(\mathbf{h}_i^L, \tilde{\mathbf{h}}_i^L\right), \tag{2.45}$$

$$\tilde{\mathbf{h}}_i^L = \sum\nolimits_{v_j \in \mathcal{N}(i)} \frac{1}{d_j} \mathbf{h}_j^L, \tag{2.46}$$

where $d_j = \sum_k \mathcal{E}(v_j, v_k)$ is the degree of node v_j, $\mathcal{D}(\cdot, \cdot)$ is a distance metric, and $\tilde{\mathbf{h}}_i^L$ is the reconstructed node representation from neighborhoods. Then, the ANIs during different gradient descend steps in the inner circle of MAML form a sequence and an LSTM is adopted to decide whether the inner optimization should be finished at each step. The controller, i.e., the LSTM, is optimized using reinforcement learning and the graph classification accuracy is adopted as rewards.

All aforementioned methods focus on specific graph tasks. G-Meta [59] proposes a more general meta-learning framework to handle various tasks. The core idea is to adopt local subgraph structures and use GNNs to learn subgraph representations. In this way, different tasks can be represented by different subgraph structures. For example, in node classification, each node can be represented by an ego-network, and in link prediction, each edge can be represented via local subgraphs for a pair of nodes. Then, a prototype architecture is leveraged where a prototype landmark $\mathbf{c}_k$ is calculated for each class k based on the subgraph representations:

$$\mathbf{c}_k = \frac{1}{N_k} \sum\nolimits_{y_{\mathcal{G}_j} = k} \mathbf{h}_{\mathcal{G}_j}, \quad N_k = \sum\nolimits_j \mathbb{1}(y_{\mathcal{G}_j} = k), \tag{2.47}$$

where $\mathbf{h}_{\mathcal{G}_j}$ and $y_{\mathcal{G}_j}$ are representations and labels for the subgraph $\mathcal{G}_j$, respectively, $\mathbb{1}(\cdot)$ is the indicator function, and N_k is the number of samples in class k. For each subgraph in both training and testing set, a class distribution vector is computed via the Euclidean distance between the subgraph representations and prototypes landmarks:

$$\mathbf{p}_{\mathcal{G}_j} = \frac{\exp(-\left\| \mathbf{h}_{\mathcal{G}_j} - \mathbf{c}_k \right\|)}{\sum_{k'} \exp(-\left\| \mathbf{h}_{\mathcal{G}_j} - \mathbf{c}_{k'} \right\|)}. \tag{2.48}$$

The architecture can be trained end-to-end via MAML since each subgraph can be considered as an independent data point.

References

1. Vanschoren, Joaquin : Meta-learning: A survey. arXiv preprint arXiv:1810.03548
2. Ravi, Sachin and Larochelle, Hugo : Optimization as a model for few-shot learning. ICLR. (2016)
3. Qiao, Siyuan and Liu, Chenxi and Shen, Wei and Yuille, Alan L : Few-shot image recognition by predicting parameters from activations. Proceedings of the IEEE Conference on Computer Vision and Pattern Recognition. 7229–7238 (2018)

 4. Gidaris, Spyros and Komodakis, Nikos : Dynamic few-shot visual learning without forgetting. Proceedings of the IEEE Conference on Computer Vision and Pattern Recognition. 4367–4375 (2018)
 5. Santoro, Adam and Bartunov, Sergey and Botvinick, Matthew and Wierstra, Daan and Lillicrap, Timothy : Meta-learning with memory-augmented neural networks. International conference on machine learning. 1842–1850 (2016)
 6. Munkhdalai, Tsendsuren and Yu, Hong : Meta networks. Proceedings of machine learning research. **70**, 2554 (2017)
 7. Graves, Alex and Wayne, Greg and Danihelka, Ivo : Neural turing machines. arXiv preprint arXiv:1410.5401
 8. Tran, Kien and Sato, Hiroshi and Kubo, Masao : Memory Augmented Matching Networks for Few-Shot Learnings. International Journal of Machine Learning and Computing. **9** 6 (2019)
 9. Jack W Rae, Sergey Bartunov, Timothy P. Lillicrap in *Meta-Learning Neural Bloom Filters*, DeepAI. https://deepai.org/publication/meta-learning-neural-bloom-filters. Cited 31 Aug 2020
10. Koch, Gregory and Zemel, Richard and Salakhutdinov, Ruslan : Siamese neural networks for one-shot image recognition. ICML deep learning workshop. **2** (2015)
11. Vinyals, Oriol and Blundell, Charles and Lillicrap, Timothy and Wierstra, Daan and others : Matching networks for one shot learning. Advances in neural information processing systems. 3630–3638 (2016)
12. Snell, Jake and Swersky, Kevin and Zemel, Richard: Prototypical networks for few-shot learning. Advances in neural information processing systems. 4077–4087 (2017)
13. Sung, Flood and Yang, Yongxin and Zhang, Li and Xiang, Tao and Torr, Philip HS and Hospedales, Timothy M : Learning to compare: Relation network for few-shot learning. Proceedings of the IEEE Conference on Computer Vision and Pattern Recognition. 1199–1208 (2018)
14. Garcia, Victor and Bruna, Joan : Few-shot learning with graph neural networks. arXiv preprint arXiv:1711.04043
15. Oreshkin, Boris and López, Pau Rodríguez and Lacoste, Alexandre : Tadam: Task dependent adaptive metric for improved few-shot learning. Advances in Neural Information Processing Systems. 721–731 (2018)
16. Ren, Mengye and Triantafillou, Eleni and Ravi, Sachin and Snell, Jake and Swersky, Kevin and Tenenbaum, Joshua B and Larochelle, Hugo and Zemel, Richard S : Meta-learning for semi-supervised few-shot classification. arXiv preprint arXiv:1803.00676
17. Allen, Kelsey R and Shelhamer, Evan and Shin, Hanul and Tenenbaum, Joshua B : Infinite mixture prototypes for few-shot learning. arXiv preprint arXiv:1902.04552
18. Li, Aoxue and Luo, Tiange and Xiang, Tao and Huang, Weiran and Wang, Liwei : Few-shot learning with global class representations. Proceedings of the IEEE International Conference on Computer Vision. 9715–9724 (2019)
19. Maclaurin, Dougal and Duvenaud, David and Adams, Ryan : Gradient-based hyperparameter optimization through reversible learning. International Conference on Machine Learning. 2113–2122 (2015)
20. Finn, Chelsea and Abbeel, Pieter and Levine, Sergey : Model-agnostic meta-learning for fast adaptation of deep networks. Proceedings of the 34th International Conference on Machine Learning. **70** 1126–1135 (2017)
21. Finn, Chelsea and Levine, Sergey : Meta-learning and universality: Deep representations and gradient descent can approximate any learning algorithm. arXiv preprint arXiv:1710.11622
22. Finn, Chelsea and Rajeswaran, Aravind and Kakade, Sham and Levine, Sergey : Online meta-learning. arXiv preprint arXiv:1902.08438
23. Rajeswaran, Aravind and Finn, Chelsea and Kakade, Sham M and Levine, Sergey : Meta-learning with implicit gradients. Advances in Neural Information Processing Systems. 113–124 (2019)
24. Hochreiter, Sepp and Younger, A Steven and Conwell, Peter R : Learning to learn using gradient descent. International Conference on Artificial Neural Networks. 87–94 (2001)

25. Andrychowicz, Marcin and Denil, Misha and Gomez, Sergio and Hoffman, Matthew W and Pfau, David and Schaul, Tom and Shillingford, Brendan and De Freitas, Nando : Learning to learn by gradient descent by gradient descent. Advances in neural information processing systems. 3981–3989 (2016)

26. Cao, Yue and Chen, Tianlong and Wang, Zhangyang and Shen, Yang : Learning to Optimize in Swarms. Advances in Neural Information Processing Systems 32. 15044–15054 (2019)

27. Li, Ke and Malik, Jitendra : Learning to optimize. arXiv preprint arXiv:1606.01885

28. Nichol, Alex and Achiam, Joshua and Schulman, John : On first-order meta-learning algorithms. arXiv preprint arXiv:1803.02999

29. Lee, Yoonho and Choi, Seungjin : Gradient-based meta-learning with learned layerwise metric and subspace. arXiv preprint arXiv:1801.05558

30. Rusu, Andrei A and Rao, Dushyant and Sygnowski, Jakub and Vinyals, Oriol and Pascanu, Razvan and Osindero, Simon and Hadsell, Raia : Meta-learning with latent embedding optimization. arXiv preprint arXiv:1807.05960

31. Grant, Erin and Finn, Chelsea and Levine, Sergey and Darrell, Trevor and Griffiths, Thomas : Recasting gradient-based meta-learning as hierarchical bayes. arXiv preprint arXiv:1801.08930

32. Finn, Chelsea and Xu, Kelvin and Levine, Sergey : Probabilistic model-agnostic meta-learning. Advances in Neural Information Processing Systems. 9516–9527 (2018)

33. Wang, Zhenyi and Zhao, Yang and Yu, Ping and Zhang, Ruiyi and Chen, Changyou : Bayesian Meta Sampling for Fast Uncertainty Adaptation. International Conference on Learning Representations. (2019)

34. Li, Junnan and Wong, Yongkang and Zhao, Qi and Kankanhalli, Mohan S : Learning to learn from noisy labeled data. Proceedings of the IEEE Conference on Computer Vision and Pattern Recognition. 5051–5059 (2019)

35. Antoniou, Antreas and Storkey, Amos J : Learning to learn by self-critique. Advances in Neural Information Processing Systems. 9940–9950 (2019)

36. Li, Zhenguo and Zhou, Fengwei and Chen, Fei and Li, Hang : Meta-sgd: Learning to learn quickly for few-shot learning. arXiv preprint arXiv:1707.09835

37. Antoniou, Antreas and Edwards, Harrison and Storkey, Amos : How to train your MAML. arXiv preprint arXiv:1810.09502

38. Park, Eunbyung and Oliva, Junier B : Meta-curvature. Advances in Neural Information Processing Systems. 3314–3324 (2019)

39. Maruan Al-Shedivat, Trapit Bansal, Yuri Burda, Ilya Sutskever, Igor Mordatch, and Pieter Abbeel. Continuous adaptation via meta-learning in nonstationary and competitive environments. *arXiv preprint arXiv:1710.03641*, 2017.

40. Chelsea Finn, Pieter Abbeel, and Sergey Levine. Model-agnostic meta-learning for fast adaptation of deep networks. *arXiv preprint arXiv:1703.03400*, 2017.

41. James Harrison, Apoorva Sharma, Chelsea Finn, and Marco Pavone. Continuous meta-learning without tasks. *arXiv preprint arXiv:1912.08866*, 2019.

42. Ricardo Vilalta and Youssef Drissi. A perspective view and survey of meta-learning. *Artificial intelligence review*, 18(2):77–95, 2002.

43. Feihu Zhang and Benjamin W Wah. Supplementary meta-learning: Towards a dynamic model for deep neural networks. In *Proceedings of the IEEE International Conference on Computer Vision*, pages 4344–4353, 2017.

44. Risto Vuorio, Shao-Hua Sun, Hexiang Hu, and Joseph J. Lim Multimodal Model-Agnostic Meta-Learning via Task-Aware Modulation *Advances in Neural Information Processing Systems*, pages 1–12, 2017.

45. Chelsea Finn, Pieter Abbeel, and Sergey Levine Model-Agnostic Meta-Learning for Fast Adaptation of Deep Networks *Proceedings of Machine Learning Research*, pages 1126–1135, 2017.

46. Chen Xing, Negar Rostamzadeh, Boris N. Oreshkin, and Pedro O. Pinheiro Adaptive Cross-Modal Few-shot Learning *Advances in Neural Information Processing Systems*, pages 4848–4858, 2019.

47. Jake Snell, Kevin Swersky, and Richard S. Zemel Prototypical Networks for Few-shot Learning *Advances in Neural Information Processing Systems*, pages 4077–4087, 2017.

48. Duan, Yan, Schulman, John, Chen, Xi, Bartlett, Peter L, Sutskever, Ilya, and Abbeel, Pieter. RI^2: Fast reinforcement learning via slow reinforcement learning. *arXiv preprint arXiv:1611.02779*, 2016.

49. Finn, Chelsea, Abbeel, Pieter, and Levine, Sergey. Model-agnostic meta-learning for fast adaptation of deep networks. *arXiv preprint arXiv:1703.03400*, 2017.

50. Hochreiter, Sepp, Younger, A Steven, and Conwell, Peter R. Learning to learn using gradient descent. In *International Conference on Artificial Neural Networks*, pp. 87–94. Springer, 2001.

51. Nichol, Alex and Schulman, John. Reptile: a scalable metalearning algorithm. *arXiv preprint arXiv:1803.02999*, 2(3):4, 2018.

52. Santoro, Adam, Bartunov, Sergey, Botvinick, Matthew, Wierstra, Daan, and Lillicrap, Timothy. Meta-learning with memory-augmented neural networks. In *International conference on machine learning*, pp. 1842–1850, 2016.

53. Wang, Jane X, Kurth-Nelson, Zeb, Tirumala, Dhruva, Soyer, Hubert, Leibo, Joel Z, Munos, Remi, Blundell, Charles, Kumaran, Dharshan, and Botvinick, Matt. Learning to reinforcement learn. *arXiv preprint arXiv:1611.05763*, 2016.

54. Avishek Joey Bose, Ankit Jain, Piero Molino, and William L Hamilton. Meta-graph: Few shot link prediction via meta learning. *arXiv preprint arXiv:1912.09867*, 2019.

55. Aaron Clauset, Mark EJ Newman, and Cristopher Moore. Finding community structure in very large networks. *Physical review E*, 70(6):066111, 2004.

56. Peng Cui, Xiao Wang, Jian Pei, and Wenwu Zhu. A survey on network embedding. *IEEE Transactions on Knowledge and Data Engineering*, 31(5):833–852, 2018.

57. Justin Gilmer, Samuel S Schoenholz, Patrick F Riley, Oriol Vinyals, and George E Dahl. Neural message passing for quantum chemistry. In *Proceedings of the 34th International Conference on Machine Learning-Volume 70*, pages 1263–1272, 2017.

58. Weihua Hu, Matthias Fey, Marinka Zitnik, Yuxiao Dong, Hongyu Ren, Bowen Liu, Michele Catasta, and Jure Leskovec. Open graph benchmark: Datasets for machine learning on graphs. *arXiv preprint arXiv:2005.00687*, 2020.

59. Kexin Huang and Marinka Zitnik. Graph meta learning via local subgraphs. *arXiv preprint arXiv:2006.07889*, 2020.

60. Ning Ma, Jiajun Bu, Jieyu Yang, Zhen Zhang, Chengwei Yao, and Zhi Yu. Few-shot graph classification with model agnostic meta-learning. *arXiv preprint arXiv:2003.08246*, 2020.

61. Jasper Snoek, Hugo Larochelle, and Ryan P Adams. Practical bayesian optimization of machine learning algorithms. In *Advances in neural information processing systems*, pages 2951–2959, 2012.

62. Ziwei Zhang, Peng Cui, and Wenwu Zhu. Deep learning on graphs: A survey. *IEEE Transactions on Knowledge and Data Engineering*, 2020.

63. Fan Zhou, Chengtai Cao, Kunpeng Zhang, Goce Trajcevski, Ting Zhong, and Ji Geng. Meta-gnn: On few-shot node classification in graph meta-learning. In *Proceedings of the 28th ACM International Conference on Information and Knowledge Management*, pages 2357–2360, 2019.

Part II
Applications of Automated Machine Learning and Meta-Learning to Multimedia

As one unique feature of this book is to introduce the impacts of AutoML and meta-learning on multimedia research, **Part II** will cover their utilizations in multimedia content including video, image and text, as well as multimedia applications ranging from search and recommendation to multimodal analysis. We note that multimedia herein is in general sense, even though natural language processing (NLP) and computer vision (CV) have their own research fields.

Chapter 3
Automated Machine Learning
for Multimedia

The term *Multimedia* has been taking on different meanings from its first advent in 1960s until today's common usage which refers *multimedia* to "an electronically delivered combination of media including videos, still images, audios, and texts in such a way that can be accessed interactively".[1] Moreover, multimedia sometimes also refers to multi-modality in machine learning community. After evolutionary development in more than two decades, multimedia research has also made great progress on image/video content analysis, multimedia search and recommendation, multimedia streaming, multimedia content delivery etc. The theory of Artificial Intelligence, a.k.a. AI, coming into the sight of academic researchers a little earlier in 1950s, has also experienced decades of development for various methodologies covering symbolic reasoning, Bayesian networks, evolutionary algorithms and all the way towards the recent deep learning technique. The wide applications of deep learning technologies promote multimedia fields with different modal forms such as text, audio, image and so on [1]. With the advent of the era of deep learning, a complex deep learning model has many configurations including hyper-parameters and neural network structure. These hyper-parameter configurations together with deep architectures are shown to have significant impacts on the performances of different deep learning models, posing a very interesting research topic of to what extend can various hyper-parameter configurations and neural architectures influence the corresponding model performances. Recently, researchers have been interested in applying AutoML frameworks such as HPO and NAS to better design and optimize real-world multimedia machine learning systems.

In this chapter, we review and summarize the application of different HPO and NAS methods for multimedia. Firstly, we will discuss scenarios of multimedia search and recommendation where people use HPO or NAS as candidates for solutions. After that, we will introduce HPO and/or NAS for different kind of modalities including text, audio, image and video.

---

[1] https://en.wikipedia.org/wiki/Multimedia

W. Zhu, X. Wang, *Automated Machine Learning and Meta-Learning for Multimedia*, https://doi.org/10.1007/978-3-030-88132-0_3

3.1 HPO for Multimedia Search and Recommendation

Nowadays, with the increasingly prominence of PC and mobile applications, the information on the web is composed of different media types, e.g., image, audio, video and text. Diversified media forms and massive data further boost two well studied and highly related application fields: multimedia search and recommendation. Meanwhile, many machine learning models are recently widely applied on famous platforms like YouTube, Facebook and Google, to improve the search and recommendation quality on multimedia content. In this section, we discuss the applications of HPO to multimedia search and recommendation.

3.1.1 Multimedia Search and Recommendation

Both concepts of *multimedia search* and *multimedia recommendation* are of quite wide scope. They play important and attractive roles in both academia and industry. Real-world multimedia search or recommendation systems are usually complex and composed of several functional modules to process the data (e.g., a multimedia search system often contains modules of indexing, query processing, search and re-ranking). However, to present the idea more clearly and intensively, in this section we only focus on the core module of search and recommendation and discuss the framework in a more academic style.

A. Multimedia Search

Multimedia search, or multimedia retrieval, is mainly related to two elements: the *query* provided by users and the *document* retrieved by the system. When the query and document are of the same media type, the multimedia search task is referred to as *query-by-example* or *query-by-humming* [59]. For instance, in the content-based image retrieval task, the query is an image and the retrieved documents are similar images on the web. The query could also be an audio recording and the system would retrieve the correct song. On the other hand, when the query and document are of different media types, the multimedia search task is referred to as *cross-modal similarity search*. A representative task is Ad-hoc Video Search (AVS) [61], where the query is a natural-language sentence and the document is the most relevant video to the sentence. Another example could be audio for image/video retrieval in the scope of music retrieval.

B. Multimedia Recommendation

Multimedia recommendation, or MultiMedia Recommender System (MMRS), includes three different tasks: Collaborative Filtering MMRS (CF-MMRS), Content-Based MMRS (CB-MMRS) and multimedia-driven RS. CF-MMRS is the same as ordinary recommender system based on community preference except that the recommended items are of multimedia types. multimedia-driven RS exploits multimedia content to give recommendation on ordinary items. In this section, we focus on CB-MMRS, which recommends multimedia items based on the multimedia content and the user history. We refer to [56] for detailed discussions about this taxonomy of MMRS.

In a CB-MMRS, the input is composed of the specific *user* (for whom we recommend multimedia items), the user's history *items* (the items he/she interacted with) and the candidate items (from which we recommend), and the output is a ranked list of the candidate items. In the scope of multimedia recommendation, the item here can be *single-modal* (visual: image or video, aural: audio, or textual: text), or *multimodal*. As typical multimodal items, micro videos or movies could be the composition of video, audio and text (subscripts). For another example, a piece of music/song could also be composed of audio, text (lyrics), image (album cover) and even video (music video).

3.1.2   Applying HPO on Multimedia Search and Recommendation

Given that HPO on multimedia search has drawn few research attentions to date, and multimedia search and multimedia recommendation may share similar model formulations under the setting of HPO, in this section we only discuss research on HPO for multimedia recommendation.

One challenging problem in recommender systems is hyper-parameter optimization (HPO). In the system design process, there are many hyper-parameters that need to be set and optimized. The massive data take a lot of time and energy for human to adjust the parameters, and the global optimal hyper-parameter setting may not be found only by experience and attempts. In order to solve these problems, HPO for recommendation aims to find a relatively optimal configuration with fewer computing resources, and greatly reduce the manual need for hyper-parameter tuning, so as to achieve end-to-end learning.

To illustrate an example of hyper-parameters in recommendation models, we first briefly introduce two latent factor models, i.e., Bias Stochastic Gradient Descent (Bias SGD) [6] and Alternating Least Squares with Weighted Regularization (ALS-WR) [7]. Both of the models construct consumer and product feature matrices in SVD form, and can be run on distributed Hadoop System [4]. Bias SGD is an improved version of SGD that additionally calculates a bias for each consumer

Table 3.1 Hyper-parameters of Bias SGD Algorithm (Chan et al. [4])

Hyper-parameter	Description
λ	Regularization to prevent overfitting
Lrate	Learning rate
Decay	Learning rate decay

Table 3.2 Hyper-parameters of ALS Algorithm (Chan et al. [4])

Hyper-parameter	Description
λ	Regularization to prevent overfitting

and product. Similarly, ALS-WR is an improved version of ALS that additionally calculates a bias for each consumer and product. The descriptions of hyper-parameters in Bias SGD and ALS algorithms are demonstrated in Tables 3.1 and 3.2 respectively. We refer interested users to the corresponding works [6, 7] for detailed explanations.

However, big data era with massive data and higher data dimensions brings many challenges to the practical application of HPO. It is difficult for people to completely verify the model in the hyper-parameter space by using traditional HPO methods such as grid search and random search. In the remaining of this section, we will provide a detailed overview of HPO for multimedia recommendation.

Chan et al. [5] present an approach to continuously re-select hyper-parameter settings of the algorithm in a large-scale retail recommender system. An automatic hyper-parameter optimization technique is applied on CF algorithms to improve prediction accuracy. They use two distributed CF algorithms, Bias SGD and ALS-WR, which can process data and construct models in parallel. For Bias SGD, they tune three hyper-parameters—regularization λ in [0.001, 0.01] to prevent overfitting, learning rate in [0.001, 0.01] and learning rate decay in [0.1, 1]. For ALS-WR, they tune regularization λ in [0.01, 0.1] to prevent overfitting. This paper mainly concerns about the comparison between the traditional one-off modeling approach (such as grid search and random search) and the continuous modeling approach.

They define HPO problem as an optimization problem written as:

$$\theta^* = \arg\min_{\theta} C(p_v, q_v, r_v, M(p_{tr}, q_{tr}, r_{tr}, \theta)), \tag{3.1}$$

and they evaluate the performance of a set of hyper-parameter settings as:

$$C(p_{test}, q_{test}, r_{test}, M(p_{tr+v}, q_{tr+v}, r_{tr+v}, \theta^*)), \tag{3.2}$$

where C is the cost function, M is the modeling function, θ is a hyper-parameter set, p_{tr} and q_{tr} are the consumer and product feature vectors of the training dataset, p_v and q_v are the consumer and product feature vectors of the validation dataset, p_{test} and q_{test} are the consumer and product feature vectors of the test dataset and r_{tr}, r_v and r_{test} are the vectors of preference scores of training, validation and test

---

Algorithm 7 Continuous hyper-parameter optimization (Chan et al. [5])

Data: D (dataset of month 1–24), A (algorithm)
Result: RMSE and MAP@k on each month of 13–24
    $trainD = D$[month 1–12];
    for $m \in \{13 \ldots 24\}$ **do**
        θ^* = select hyper-parameters of A based on $trainD$;
        P = model of A trained by $trainD$ and θ^*;
        evaluate RMSE of P on D[m];
        evaluate MAP@k of P on D[m];
        $trainD = trainD + D$[m];
    end for

---

datasets respectively. The approach to search for better hyper-parameters before each iteration of model re-training with the addition of new data is described in Algorithm 7 as follows.

Different from HPO methods such as grid search and random search, Chan et al. [5] consider the effect of re-selecting new hyper-parameters after new data are collected over time. However, simply re-training the model with the addition of new data may not be sufficient to accommodate the frequently and rapidly changes of business inputs. Thus, they put forward the hypothesis that a recommender system model may render less accuracy, or even become invalid, as the business dynamic changes over time. Under this hypothesis, a continuous HPO approach can improve prediction accuracy of a recommender system in a retail environment. They investigate whether re-selecting hyper-parameters for model selection repeatedly after the introduction of new data would have any impact on the prediction accuracy of recommendation systems. The dataset is provided by a large UK retail chain business, which is a 2-year anonymized product purchase records of loyalty card holders on the retailer's e-commerce site and in all physical stores of the retailer in the UK. It contains complete transaction records of 10,217,972 unique loyalty card holders and 2939 unique products under 10 selected brands. There are also 21,668,137 in-store purchase transaction records and 2,583,531 online purchase transaction records. The result shows their model improves prediction accuracy of a recommendation system for both online e-commerce site and offline retail stores of the retail chain business. This paper presents a new direction for improving the predictive performance of large recommendation systems in real retail scenarios.

Matuszyk et al. [8] investigate the applicability of nine different optimization strategies to recommendation systems, including random walk, genetic algorithm, sequential model-based algorithm configuration, full enumeration, random search, greedy search, simulated annealing, Nelder–Mead and particle swarm optimization.

The search space is defined with three dimensions. The first hyper-parameter is the number of latent dimensions k from the matrix factorization algorithm in the range [10, 200]. The second hyper-parameter is η in the range [0.001, 0.1]. It is learning rate used by the SGD in the process of factorizing a rating matrix. The third is a regularization parameter λ used also by the SGD to prevent overly high latent factors in the matrix factorization range in [0.001, 0.1].

Algorithm 8 Evaluation framework (Matuszyk et al. [8])

Input: $\mathcal{HO}$ (list of hyper-parameter optimizer)
     X (dataset)
     RS (recommender system)
     M (number of repetitions, e.g., 100)
     N (number of experiments, e.g., 50)
for $i \in \{1, \cdots, M\}$ **do**
    for $j \in \{1, \cdots, N\}$ **do**
      $(X_{RST_r}, X_{RST_e}, X_{eval}) \leftarrow randSplit(X);$
      for $HO \in \mathcal{HO}$ **do**
        /*   Training phase   */
        $\theta^* \leftarrow nextParam(HO, res_{HO});$
        $re_{tr} \leftarrow trainRS(RS, \theta, X_{RST_e});$
        $perf_{tr} \leftarrow evalPerf(rs_{tr}, X_{RST_e});$
        $res_{HO} \leftarrow res_{HO} \bigcup \{(\theta, perf_{tr}\};$
        /*   Evaluation phase   */
        $\theta^* \leftarrow getBestParam(res_{HO});$
        $rs_{ev} \leftarrow trainRS(RS, \theta^*, X_{RST_r} \bigcup X_{RST_e});$
        $perf_{ev}^{(i,HO,j)} \leftarrow evalPerf(rs_{ev}, X_{eval})$
      end for
    end for
end for

They use real-world datasets MovisLens 1M and 100k to evaluate different hyper-parameter optimization methods. Their experiments show that random search outperforms random walk, greedy algorithm, and grid search on all datasets or, in the best case, converged to the level of random search. SMAC, PSO and genetic algorithms performed similarly to the random search. Because of many advantages of random search, such as full parallelization, simplicity, constant and nearly negligible computation time, random search is recommended for optimizing hyper-parameters in the domain of recommender systems. Only in application scenarios, where marginal improvements play an important role the Nelder–Mead algorithm or of simulated annealing is a good choice because the parameters are numerical and parallelization is not necessary.

Online news is a media for people to get new information quickly and conveniently. There are a lot of online news media out there but many people will only read news that is interesting for them. This kind of news tends to be popular and will bring profit to the media owner. That's why it is necessary to predict whether a news is popular or not by using the prediction methods. Wicaksono et al. [9] use genetic algorithm to get optimal hyper-parameter in predicting whether some news is popular or not. The experiment is implemented using Scikit Learn library and the data used is the online news dataset downloaded from UCI machine learning site. Each of the machine learning has different hyper-parameter to optimize. In SVM, the chromosomes of genetic algorithm will be penalty coefficient of objective function C, coefficient of kernel function γ, and kernel. In AdaBoost, the chromosomes are the number of estimator and learning rate. In random forest, the parameters are the number of decision tree used, minimum sample leaf, and minimum weight fraction of leaf. Finally, the number of K is the hyper-parameter to optimize in KNN method. The crossover rate used is 0.5 and the mutation rate is 0.1. The tournament will be used as evaluation method to determine the next generation of the population. Lastly, the number of iterations used is 10. To implement this method, they use evolutionary algorithm search in Scikit-learn with population size 50 and generation number 1000. Figure 3.1 demonstrates the diagram of the prediction algorithm.

Caselles et al. [10] investigate the marginal importance of hyper-parameters in a recommendation setting through large hyper-parameter grid search on various datasets. Specifically, they use Next Event Prediction (NEP) as an offline proxy and a common kind of Word2vec (W2V) method Skip-Gram with Negative Sampling (SGNS). They perform grid searches on four different types of recommendation datasets (two of music, one of e-commerce and one of click-stream). They choose seven hyper-parameters to be optimized: the number of epochs n (10 to 200 with step of +10), the window-size L (3, 7, 12, 15), the sub-sampling parameter t (10^{-5} to 10^{-1} with step of $\times 10$), the negative sampling distribution parameter α (-1.4 to 1.4 with step of +0.2), the embedding size (50 to 200 with a step of 50), the number of negative samples (5 to 20 with a step of 5) and the learning rate (0.0025 to 0.25 with a step of $\times 10$). Because the marginal benefit of the 3 latter variables to the optimization is less than 2% in terms of performance, they tune four hyper-parameters, namely negative sampling distribution, number of epochs, subsampling parameter and window-size and keep the other fixed to default values. The Result reveals that optimizing some neglected hyper-parameters improves performance on a recommendation task, and the choices for hyper-parameters are data and task dependent. Comparing several recommendation datasets on the same task, they observe that different data distributions result in different optimal hyper-parameter values.

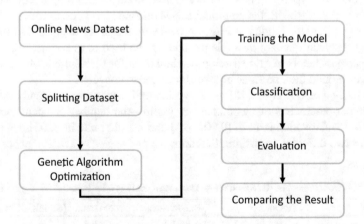

Fig. 3.1 Diagram of predicting whether a news is popular

Bayesian hyper-parameter optimization is a model-based sequential hyper-parameter optimization that hyper-parameters can be selected based on the evaluation of past models. It has been widely used in the tuning of various machine learning algorithms. Bayesian optimization assumes there is a functional relationship $f(x)$ between the hyper-parameter and an evaluation index that the algorithm needs to optimize, where x is in a bounded set $x \in \mathbb{R}^D$. Bayesian optimization establishes a probabilistic model for $f(x)$, and explore the model to determine the next assessment point in χ. Dewancker et al. [11] use a variant of collaborative filtering known as quadratically regularized PCA [13]. This method aims to learn latent factors X, Y that best recreate the ratings matrix A, with a regularization penalty coefficient λ on the learned factors.

As mentioned above, quadratically regularized PCA aims to learn latent factors X, Y that best recreate the ratings matrix A:

$$\text{argmin}_{x_i, y_j} \sum_{i=1}^{m} \sum_{j=1}^{n} (A_{ij} - x_i y_i)^2 + \lambda \sum_{i=1}^{m} ||x_i||_2^2 + \lambda \sum_{j=1}^{m} ||y_j||_2^2, \qquad (3.3)$$

where x_i represents the i-th row of the X factor matrix and y_j represents the j-th column of the Y factor matrix. Let λ denotes the regularization term, k denotes the desired rank of the factorization, and T denotes the number of iterations of each alternating step in the minimization as tunable parameters to SigOpt. In this experiment they consider the MovieLens dataset and use the MLlib package within Apache Spark. The code for this example is available in the SigOpt examples github repository. SigOpt tunes the alternating least square algorithm parameters with respect to the RMSE of the validation set. They compare the RMSE on the test ratings after tuning ALS algorithm between SigOpt, Random Search and Default MLlib ALS with no tuning. The results in Table 3.3 show that SigOpt works best.

Dey et al. [14] provide a formulation of empirical Bayes described by Atchade [12] to tune the hyper-parameters of priors used in Bayesian set up of collaborative filter. In terms of RMSE, the empirical Bayes method is as good as the result using straight grid search. Empirical Bayes provides much faster computation than the usual grid search method, thus this method can be used to guess an initial choice for hyper-parameters in grid search procedure even for the datasets where MCMC oscillates around the true value or takes long time to converge.

Similarly, Galuzzi et al. [15] use Bayesian optimization to efficiently optimize three hyper-parameters of recommender system—the number of latent factors in [10, 50], regularization term in [0.001, 0.1] and learning rate in [0.001, 0.1]. They consider two different acquisition functions: EI and ϵ-greedy TS. The paper shows

Table 3.3 Comparison of RMSE on the hold out (test) ratings after tuning ALS algorithm [11]

Model	SigOpt	Random search	No tuning (default MLlib ALS)
Hold out RMSE	0.7864 (−40.7%)	0.7901	1.3263

how HPO for a collaborative filtering based recommender systems can be efficiently performed through Bayesian optimization.

Gautam et al. [16] use parameter tuning in a huge music catalog repository across multiple languages. Because of their effort in increasing the ability of discovery in an unsupervised setting, they do not use Akaike Information Criterion (AIC) and the Bayesian Information Criterion (BIC), both of which try to select inductive bias by penalizing model fit. There is a wide variety of features to solve the problem, so it becomes essential to select a model which is tuned according to their use cases. This could be achieved through conducting a rigorous set of experiments for final model selections. There are indeed a set of parameters playing important roles for model selections. They consider three activation functions including ReLu, sigmoid and tanh, iterations, number of clusters, number of layers in deep neural network, number of hidden neurons/filters in each layer of the network. All above mentioned parameters are considered while training deep neural networks.

Tuning hyper-parameters for large-scale matrix factorization (MF) is very time consuming. Some scholars solve the HPO problem of recommendation system from another angle. To solve this problem, Zeng et al. [17] propose a scale-invariant parametric matrix factorization method named FAVA, where a set of scale-invariant parameters are defined for model complexity regularization. Different from traditional matrix factorization methods, this method considers the time costs when tuning hyper-parameters for large-scale MF problems. They tune the regularization λ from 0.1 to 4.0 by step size 0.1 and $\gamma \in \{$ 10, 20, 30, 40, 50, 60, 70, 80, 90, 100, 200, 300, 400, 500, 600, 700, 800, 900, 1000, 2000, 3000, 4000, 5000, 6000, 7000, 8000, 9000, 10000$\}$, where λ and γ are hyper-parameters in convex formulation that can estimate the factorization variances. The trade-off form [18, 19] can be expressed as:

$$min_X F(X) = \sum_{(i,j) \in \Omega} (R_{i,j} - X_{i,j})^2 + \lambda tr(X), \tag{3.4}$$

$$s.t. X \succeq 0. \tag{3.5}$$

The trace bounding form [20, 21] can be written as:

$$min_X G(X) = \sum_{(i,j) \in \Omega} (R_{i,j} - X_{i,j})^2, \tag{3.6}$$

$$s.t. X \succeq 0, \tag{3.7}$$

$$s.t. tr(X) \leq y. \tag{3.8}$$

Because the optimal hyper-parameters usually change with the different scale of matrices, they propose a scale-invariant parametric MF method. Thus, they can find the best hyper-parameters λ^* on small sub-matrix and then directly exploit it into the original large matrix. The proposed method can thus free people from tuning hyper-parameters on large-scale matrix and achieve good performance in a more efficient manner.

3.2 HPO for Computer Vision

In this section, we will discuss the growing and exciting literature on HPO for different modalities (such as image, video, audio and text) in computer vision where there are multiple tasks, with problem formulation and method changes across the tasks. However, a common thread underlying all tasks is still similar to previous chapters. Most of the existing literature adopt random search, genetic algorithm, Bayesian optimization, Dynamic Particles Warm Optimization (DPSO) and their improved algorithms.

3.2.1 HPO for Image Applications

Due to the continuous development of science and technology, the quantity and carrier of information have undergone tremendous changes. Image has gradually become one of the main ways of information transmission because it can directly express large amount of information. The amount of image data also increases rapidly. People need a fast, efficient and reasonable way to process, analyze and interpret image data, so as to extract the valuable information efficiently and accurately from the massive image data. In this section, we will discuss applications of HPO for solving image applications including image classification and medical image recognition.

Neural network has a very good performance in the field of image classification. However, it is difficult to determine the network structure, select the network training parameters and set the network weight when we are designing a neural network. The parameters and the structure of neural network greatly affect the performance of a neural network. The problem is that the design of neural network needs a lot of human intervention, which is time-consuming and laborious, and the optimal solution may not be obtained.

The parameters and the structure of neural network greatly affect the performance of a neural network. Since the prevalence of AutoML, more and more methods for automatic optimization of neural network structure have been proposed in the hope that neural networks with better performance can be automatically and quickly searched. We note that neural architecture search (NAS) aims to automatically search neural network architecture with the help of machine learning algorithm. After the search space is defined, the new network structure is generated based on the search strategy selected in the algorithm, and finally the performance of the new network structure is evaluated. The search space of neural network structure includes the number of network layers, the size of convolution kernel and the operator of each layer. Although some researchers also consider some configurations of the network structure as hyper-parameters, we introduce HPO and NAS separately. We discuss the optimization process, the handling of the data and the architecture separately. In this section, we introduce HPO used in neural networks for images. Generally, people divide the hyper-parameters into three types: real, integer and categorical. Let us first begin with a quick review of HPO for Deep Neural Networks (DNNs) which will be helpful to the content of this section.

After building a network architecture, people use different optimizers to optimize loss function of the network. A good optimization algorithm can often train the internal parameters of the network efficiently and accurately. Here we display the hyper-parameters of four selectable optimizers introduced by Lakhmiri et al. [43]:

- Stochastic Gradient Descent (SGD): initial learning rate, momentum, dampening, weight dacey.
- Adam: initial learning rate, β_1, β_2, weight decay.
- Adagrad: initial learning rate, learning rate decay, initial accumulator, weight decay.
- RMSProp: initial learning rate, momentum, α, weight decay.

The type of hyper-parameters above is real and the scope of the hyper-parameters below is [0, 1]. In total, the optimizer of the network relies on five hyper-parameters including one categorical hyper-parameter that determines which optimizer is chosen.

Before training a network, the data is always separated into training set, testing set and validation set. Because a network needs to be repeated many times to fit and converge, it is not enough to train all the data once. How to feed the training set into network is also very important. In the actual training, all data is divided into batches and is sent in batches. For example, we can input the data all at once, one by one or input the mini-batches of data. The size of mini-batches is an integer hyper-parameter that varies between [1, n_{train}], where n_{train} is the size of training data.

Each time, when all the data is sent into the network to complete a forward calculation and a backward propagation process, we called the process an epoch. The setting of the epoch is also important. Generally speaking, the size of a good epoch is related to the degree of diversification of the data set. The greater the degree of diversification, the larger epoch should be selected. If the number of epochs is too small, the network might have been under-fitting (i.e., not learning enough about given data). If there are too many epochs, an over-fitting might occur. The early stop method is designed to solve the problem that the number of epochs needed to be set manually. It can also be considered as a regularization method to avoid over-fitting of the network. After the end of each epoch (or after N epochs), test results are obtained on the validation set. With the increase of epochs, if test errors are found to increase on the validation set, then training is stopped. The weight after the stop is taken as the final parameter of the network.

A. Image Classification

Lakhmiri et al. [43] introduce- the HyperNOMAD package that applies the MADS algorithm to simultaneously tune the hyper-parameters responsible for both the architecture and the learning process of a DNN. The workflow of HyperNOMAD is shown in Fig. 3.2 and the choices of optimizers as well as the corresponding hyper-parameters are presented in Table 3.4.

The HyperNOMAD package is a framework package for hyper-parameter optimization of DNNs using the NOMAD software. It is used to optimize a black box responsible for constructing, training a neural network and evaluating the test accuracy depending on the values of the hyper-parameters. The NOMAD software [44] is a C++ implementation of the MADS algorithm. Each iteration of MADS is divided into two steps: The search and the poll. The search phase is optional and can contain different strategies to explore a wider space in order to generate a finite number of possible mesh candidates. This step can be based on surrogate functions, Latin hyper-cube sampling, etc. The poll is strictly defined since the convergence theory of MADS relies solely on this phase. In addition, NOMAD can handle categorical variables by adding a step in the basic MADS algorithm. For more detail on how MADS handles categorical variables, the reader is referred to [44]. Lakhmiri et al. apply HyperNOMAD to the MNIST [45] and CIFAR-10 [46]

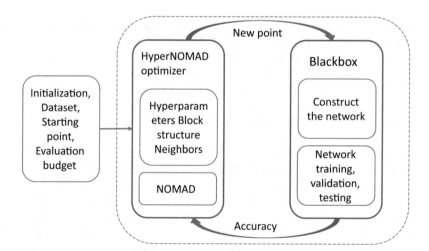

Fig. 3.2 The HyperNOMAD workflow

Table 3.4 Choices of the optimizers and the corresponding hyper-parameters [43]

Algorithm	Hyper-parameters
SVM	Bias, cost parameter, and regularization parameter
Boosted regression trees	Feature sampling rate, data sampling rate, learning rate, # trees, # leaves, and minimum # instance per leaf

datasets and compare it with other methods. The framework finds better solutions than TPE and a random search.

When developing networks for new applications, it takes a lot of time to search for a good configuration. Bochinski et al. [42] propose an evolutionary algorithm-based (EA-based) framework to automatically optimize the CNN structure by means of hyper-parameters. Further, they extend the EA-based hyper-parameter optimization towards a joint optimization of a committee of CNNs to leverage specialization and cooperation among the individual networks.

B. Medical Image Recognition

HPO can also be used in medical multimedia. Borgli et al. [53] optimize hyper-parameters of CNN models utilized in computer-aided diagnosis (CAD) systems. These systems assist physicians in the diagnosis of diseases and anomalies using visual, textual or sensory data from endoscopic examinations. In the field of gastroenterology, they use CADs to apply machine learning on the video streams for automatic detection and classification of anomalies, diseases, and medical procedures. The video streams are collected using a camera that is inserted into body cavities such as colon and esophagus. They use a pre-trained CNN model, which is fine-tuned to datasets containing images from the GastroIntestinal (GI) tract. The Bayesian optimization procedure in their system and the system flow are shown in Figs. 3.3 and 3.4, respectively.

The presented system utilizes Bayesian optimization and is used to present experiments with three automatic HPO strategies for CNN models on two gastrointestinal datasets. They focus on four hyper-parameters: The pre-trained model, the gradient descent optimizing function, the learning rate and the delimiting layer. The following gradient descent optimizer available in Keras are used: SGD, RMSprop,

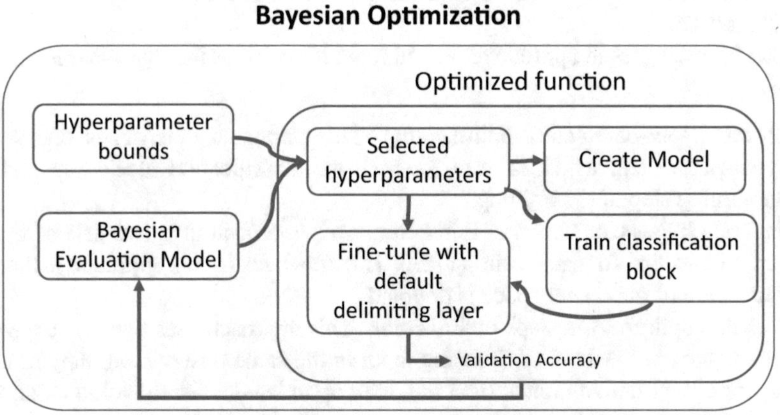

Fig. 3.3 Bayesian optimization used in Borgli et al.'s work [53]

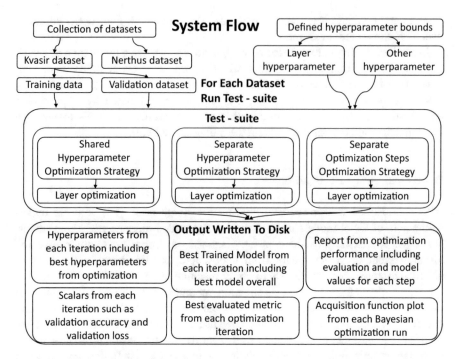

Fig. 3.4 The system flow adopted in [53]

Adagrad, Adadelta, Adam, Adamax, and Nadam. They also use these pre-trained models available in Keras: Xception, VGG16, VGG19, ResNet50, InceptionV3, InceptionResNetV2, DenseNet121, DenseNet169, and DenseNet201. The learning rate is set continuously between 1 and 10^{-4}. The delimiting layer is set between 0 and the number of layers in the chosen model. Before performing the layer optimization, they use a default delimiting layer of 2/3 of the number of layers in the model.

The following three approaches are adopted to optimize the hyper-parameters of CNN:

- Shared hyper-parameters optimization. This approach determines one set of hyper-parameters which is used for both the classification block step and the fine-tuning step of the training.
- Separate hyper-parameters optimization. This approach uses two sets of hyper-parameters, one for each training step. The drawback to this approach is that the dimension of parameter space is doubled.
- Separate optimization steps optimization. This approach uses two sets of hyper-parameters, but instead of choosing them in the same observation, they run two separate Bayesian optimization. First, they optimize the classification block, and then the best model is used for separate optimization of the fine-tuning.

The results indicate that only the shared hyper-parameters approach is successful. For the shared hyper-parameters approach, the results for both experiments show an increase of about 10% over the best approaches which they use for comparison. For similar transfer learning approaches, the difference is even greater. Therefore, it can be concluded that automatic HPO is an effective strategy for increasing performance in transfer learning cases.

3.2.2 HPO for Video Applications

In this part we review applications of HPO for solving some real-world problems for videos including face recognition, emotion recognition, object classification, dangerous tools detection and human gender classification. In the literature, HPO serves as a part of complex video-based systems.

A. Face Recognition

In the information age, biometric technology is a convenient and safe way of identity authentication, which utilizes the body's inherent characteristics such as fingerprints and facial features to do personal identification by means of computer technology. HPO can be used to optimize the performance of biometric recognition systems. Connolly et al. [47] propose an adaptive classification system (ACS) for video-based face recognition in order to reduce the damage to previous knowledge when learning new data, as shown in Fig. 3.5. In this work, they use dynamic particleswarm optimization (DPSO) algorithm [3] to adjust the hyper-parameters adaptively, which can find and track multiple local optimal solutions in the optimization space.

When incremental learning is carried out in a dynamic environment, the classifier should not only retain the previously acquired knowledge but also adjust its hyper-parameters to adapt to the new data set characteristics. So, they define their incremental learning problem as a dynamic optimization problem in a hyper-parameter space as shown in Fig. 3.6. They combine the fuzzy ARTMAP neural network classifier [2] which is suitable for incremental learning with the DPSO algorithm. They also propose a novel DPSO-based learning strategy that optimizes classifier weights, architecture, and user-defined hyper-parameters simultaneously.

They use the real-world video database to verify the impact of supervised incremental learning and dynamic hyper-parameter optimization on system performance. The experimental results show that in the process of incremental learning, optimizing the hyper-parameters of fuzzy ARTMAP can achieve a better classification rate. Furthermore, their second experimental result shows that the peak value of the objective function (in the hyper-parameter space) changes with new samples; If the dynamic optimization algorithm is not used to adjust the classifier hyper-parameters, the classification rate of fuzzy ARTMAP will decrease. Finally, they find that the optimization of ACS parameters is an expensive process that requires

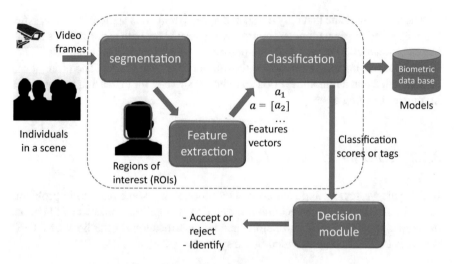

Fig. 3.5 A general biometric system for face recognition where both classification module and biometric database are replaced by the adaptive classification system

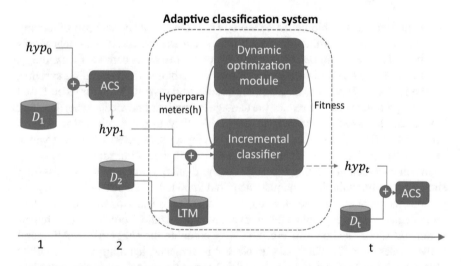

Fig. 3.6 The evolution of a new adaptive classification system (ACS) according to generic incremental learning scenario. New blocks of data are used by the ACS to update the classifier over time. Let $D_1, D_2, \cdots$ be blocks of learning data available at different instants in time. The ACS starts with an initial hypothesis hyp_0 which constitutes the prior knowledge of the domain. Each hypothesis hyp_{t-1} are updated to hyp_t by the ACS on the basis of the new data block D_t

a lot of resources, so they think fitness-based tests should be designed to identify in which situation ACS would benefit from incremental learning blocks of blocks of data in future work.

B. Emotion Recognition

Kahou et al. [48] use a support vector machine (SVM) for classification in their emotion recognition in the wild (EmotiW) Challenge. Before final classification, they combine different modality-specific models in order to learn features from multiple modalities for label assignment. They consider the hyper-parameters correspond to a kernel width term, coefficient of kernel function γ and the penalty coefficient of objective function c of SVM. At first, they simply concatenat the results of their ConvNet 1 and audio model using vectors shown in Fig. 3.7 and roughly search hyper-parameters of the SVM twice. Then, they do fine grid search over integer powers of 10 and non-integer powers of 2 within the reduced region of space. This process yields an accuracy of 42.17% on the validation set, producing a test accuracy of 38.46%.

However, when they are using the predictions of all models, the HPO process above quickly overfits. Thus, they use a more sophisticated SVM hyper-parameter search to re-weight different models and their predictions for different emotions. They implemented this via a search over discretized [0, 1, 2, 3] per dimension scaling factors. While this results in 28 additional hyper-parameters, this discretization strategy allowed them to explore all combinations. This more detailed hyper-parameter tuning increases the validation set performance to 43.69%, and the test set performance to 32.69%.

They also use random search for weighting models. Random search is an effective strategy for HPO even when the dimension of hyper-parameter space number in the tens. To perform the random search, they first sampled random weights from a uniform distribution and then normalized them to produce seven simplexes. After running this sampling procedure they get validation set performance 47.98% and test set accuracy of 39.42%. They use the results of this initial random search to initiate a local search procedure, then use fine grid search for SVM. In this procedure, they generate random weights using a Gaussian distribution around the

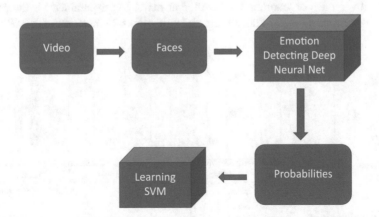

Fig. 3.7 Complete pipeline describing the final strategy used for ConvNet 1 model

best weights found so far. The weights are tested by calculating the accuracy of the so-weighted average predictions on the validation set. This strategy yielded 41.03% after a long-time search.

C. Object Classification

Yaseen et al. [49] build a object classification system from multiple videos. The CNN archecture is shown in Fig. 3.8. They propose an HPO based strategy using a mathematical model to achieve higher object classification accuracy. The mathematical model help to observe the hyper-parameter outcomes on overall performance of the learned model. Although there are many HPO methods such as racing algorithms, gradient search, Gaussian process model and Tree Parzen Estimator, these methods are hampered because of high computation requirement and only perform well for problems with a few numerical hyper-parameters. So they choose to do HPO manually that is less resource intensive and consumes less time as compared to automated methods. The evaluation of a poor hyper-parameter setting can be quickly detected by human operators after a few steps of the stochastic gradient descent algorithm. They can quickly judge if the network is performing bad and then terminate the evaluation.

They provide an analysis of these hyper-parameters and present the optimal tuning hyper-parameters. They show that the proposed system performs object classification with high accuracy and they demonstrate experimentally that the distributed training with iterative reduce for automatic video analytic is a promising way of speeding up the training process. After training, the classifier can be stored locally and uses a match probability to classify objects.

The parameters are tracked and represented in the form of graphs over multiple time stamps in order to observe the trend in the behavior of the system. The authors also present visualizations to analyze the trend in the loss function. The learning rates vary with many different values and le^{-2} is shown to be the best for the divergence of learning curve. Another major hyper-parameter is the proper normalization of the data. The authors show that the L2 employed with SGD, i.e.,

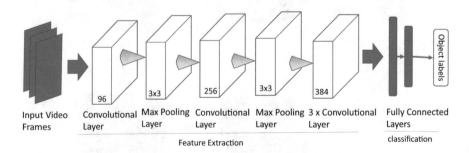

Fig. 3.8 Schematic diagram of the proposed CNN for object classification

$W_{t+1} = W_t - \alpha \delta L(\theta_t)$, is a good scheme where α varies from le^{-2}, le^{-4} to le^{-6}. They also analyze the influence of different layer activations in the same way as normalization where the regularization scheme i.e., $\lambda_2 \sum_i \theta_i^2$, is well adopted.

In the future, they plan to develop a rule-based recommendation system for cloud-based video analytic which will provide recommendations for hyperparameter tuning on the basis of input dataset and its characteristics.

D. Dangerous Tools Detection

Krol et al. [50] investigate HPO in video stream from Closed Circuit Video Television (CCTV). The system aims to find out and flash a warning if a person carrying dangerous tools pass by. Thus, the target is to have the maximum possible opportunity to detect dangerous tools promised on low number of false positives. The system consists of two modules, one is the neural network module (denoted as H_NN) and the other is the decision maker module (denoted as H_DM). H_NN is a deep neural network trained utilizing intra-frame information that selected using a sliding window of a constant size. And it returns the coordinates of the sub-image with highest probability of containing a dangerous object and the probability values. H_DM utilizes the inter-frame information that the dangerous object is usually visible on multiple consecutive frames and should be located in generally similar location on the consecutive frames. It returns coordinates of detection (or (-1, -1) value when no detection has been determined).

In the research, they do experiments with numerous (over 40) different architectures on the H_NN module, each requiring a separate set of H_DM parameters. H_DM requires three parameters:

- $Maximumdetection$ represents the amount of subsequent frames in the video that contain positive detections. Range in [2,150].
- $Maximumdistance$ is the distance between sub-images on subsequent frames with detections measured in pixels. This parameter utilizes assumption, that a dangerous tool should not move far at recording speed of 25 Frames Per Second (FPS). Range in [2,150].
- $Threshold$ means the minimum probability that given sub-image contains firearms. It defines the sensitivity of the H_DM module. Range in [0,100].

The experiments are conducted on random search, genetic algorithm, simulated annealing and Bayesian optimization. They are tested in terms of their efficiency and effectiveness in order to find the best one for the HEIMDAL project. The results presented in Table 3.5 shows Bayesian optimization provides score comparable to genetic algorithm in 20 times shorter time. The simulated annealing algorithm provides the worst average score. In the end they choose for genetic algorithm as it gives the best and the most stable result omitting time performance.

Table 3.5 Comparison of algorithms in terms of time and score on the training set, taken from [50]

Algorithm	Average execution time (s)	Average score	Standard deviation of score
Random search	69.3	15.7	3.08
Boosted regression trees	2540.3	23.4	2.02
Simulated annealing	52.2	14.3	8.34
Bayesian optimization	96.8	22.2	3.14

Algorithm 9 Nelder–Mead algorithm [51]

Initialization: Choose an initial simplex of vertices
$Y_0 = \{y_0^0, y_0^1, \cdots, y_0^n\}$. Evaluate f at the points in Y_0.
Choose constants: $0 < \gamma^s < 1, -1 < \delta^{ic} < 0 < \delta^{oc} < \delta^r, \delta^e$.
    for k = 0,1,$\cdots$ **do**
        Set $Y = Y_k$;
        Order;
        Reflect;
        Expand;
        Contract;
        Shrink;
    end for

E. Human Gender Classification

Nowadays, understanding demographic information on social influence events is important for target customer analysis. People always use human gender classification and classify age group to monitor the crowded event. Htet et al. [51] use Nelder–Mead method to tuning the performance of classification task can be defined as recognition rate, i.e., percentage of correctly classified images. Nelder–Mead algorithm is an algorithm to find the local minimum value of function of multiple variables. The advantage is that the function does not need to be differentiable and can converge to the local minimum quickly. This paper proposes new HPO configuring model for Myanmar people gender classification with competitive performance on the Myanmar image dataset and Asia image dataset.

3.2.3   HPO for Audio Applications

Many audio processing tasks such as source separation, denoising and compression need realistic and flexible models that capture the physical properties of audio signals. However, constructing useful models that are physically realistic, generic, flexible, and computationally light at the same time is an art.

Dikmen et al. [52] review three methods that deal with HPO of Gamma Markov Random Fields (GMRFs): pseudolikelihood, score matching, and contrastive divergence. They optimize the hyper-parameters of their GMRF-based audio model using contrastive divergence and compare this method to alternatives.

Deng et al. [35] provide an overview of the invited and contributed paper presented at the special session at ICASSP 2013. Five ways of improving deep learning methods are introduced, including better ways to determine the hyper-parameters of Deep Neural Networks (DNN). The performance of DNNs is sensitive to hyper-parameters such as learning rate, the strength of the regularizer, the number of layers and the number of units per layer. Sensible values for one hyper-parameter may depend on the values chosen for other hyper-parameters. Hyper-parameter tuning in DNNs is especially expensive because testing a single setting of the hyper-parameters is costly. Papers in this special session describe two methods for tackling this problem: paper [36] uses an off-the-shelf Bayesian optimization procedure [25], while paper [37] employs a sampling procedure [38] to avoid the expense of a full grid search.

Raffel et al. [39] use Bayesian optimization to optimize the design of Dynamic time warping (DTW) based alignment and matching systems. They use DTW to determine a correspondence between discrete times in the audio and MIDI files. DTW uses dynamic programming to find a monotonic alignment such that the sum of a distance-like cost between aligned feature vectors is minimized. The present work aims to remedy this by searching across a large space of DTW designs to optimize both alignment accuracy and confidence reporting. They propose a method for creating a synthetic dataset of MIDI-audio pairs by applying realistic corruptions to MIDI files, allowing them to know the correct alignment in advance. They then tune hyper-parameters for alignment using Bayesian optimization and for confidence reporting using an exhaustive search. Finally, they perform a large-scale qualitative evaluation of proposed alignment system on real-world data and discuss possibilities for improvement.

They choose to optimize over the following parameter space:

- Feature representation: They compute log-magnitude features in both chroma vectors and constant-Q spectra according to preliminary experiments.
- Time scale: Choose computed feature vectors every 46 ms or utilized beat-synchronous features.
- Normalization: They optionally z-score the feature vectors, and normalize them by their L^1, L^2, L^∞ norm, or they are not normalized at all.
- Penalty: They optimize penalty in [0, 3] to apply to the median distance between all pairs of feature vectors in X and Y. Using the median distance made this penalty adaptive to different feature representations and normalization schemes.
- Gully and band path constraint: They allowed the gully to take any value in [0, 1] and optionally enforced the band path constraint.

To evaluate the applicability of the system, they perform a large-scale qualitative evaluation on real-world data that randomly contain 500 MIDI/audio pairs from the 25,000-pair "clean MIDI datase" [40] obtained by matching MIDI files with reliable metadata to entries in the Million Song Dataset [41]. Overall, their "gold standard" alignment system successfully aligns most correctly-matched pairs and produce a reliable confidence score. In summary, large-scale optimization over synthetic data has delivered a DTW-based system which is simple to implement and achieves accurate and reliable results for both alignment and matching.

3.3 HPO for Natural Language Processing

Natural Language Processing (NLP) is a subfield of artificial intelligence (AI), which is designed to make computers understand human language. It is a cross field between computer science and linguistics. NLP explores how the human language is processed and transformed into machine language, involving the processes of cognition, understanding and generation. *Natural language cognition* and *natural language understanding* aim to make the computer turn the natural language into meaningful symbols and establish corresponding relationships, then process them according to practical purposes. *Natural language generation* generates natural language from a machine representation system such as a knowledge base or logical form. The main tasks of NLP are to study the models that are strong to represent language and to establish and implement the computing framework. After constantly improving the model, an NLP system realizes the communication between natural language and computers.

In the early stage of NLP, scientists proposed a grammatical-based NLP approach that requires complex grammatical rules, huge computational power and time. Grammatical-based NLP also has difficulty in dealing with word with various meanings. In the 1970s, scientists came up with a statistics-based approach, which builds mathematical models based on the contextual characteristics of natural language. Statistics-based NLP can judge whether a sentence is reasonable according to the possibility of the existence of a sentence. It can judge the meaning of a word in the sentence by calculating the word with the highest probability of occurrence of the position in this sentence. From then on, the problem of NLP is transformed into a mathematical problem. Nowadays, most of NLP strategies are based on statistical methods, widely used in machine translation, speech recognition, printing or handwriting recognition, spelling correction and Chinese character input. Deep learning also has been widely used in NLP because of its powerful computing power.

NLP process can be divided into six steps:

Step 1   Obtain corpus through existing data, public data sets, crawler crawling and other ways.

Step 2   Preprocess the corpus, including corpus cleaning, word segmentation, part of speech tagging, removing inessential words and so on.

Step 3   Feature engineering, namely vectorization, which represents the segmentation and the words after segmentation as the computing types that can be recognized by the computer to better express the similarity between different words.

Step 4   Feature selection to select appropriate and expressive features for the next
step of training.
Step 5   Model training, including traditional supervised, semi-supervised and
unsupervised learning models, which can be selected according to different
application needs.
Step 6   Evaluate the effect after modeling. Common evaluation indexes include
Precision, Recall, f-measure, etc.

However, overfitting and underfitting may occur in the training model. Overfitting refers to learning too much noise from the data characteristics, while
underfitting refers to the data that cannot be well fitted. The methods to solve
overfitting mainly include increasing regularization items so as to increase the
training amount of data. To solve underfitting problem, regularization items should
be reduced and other feature items should be added to process data. In the sixth step,
we always use evaluation indexes including precision, recall, F-value, etc. Accuracy
is a measure of the precision of the retrieval system. Recall rate is a measure of
the recall rate of the retrieval system. The F-value is an indicator for the overall
accuracy rate and recall rate. When the F-value is high, it indicates that the test
method is effective.

3.3.1   Bayesian Optimization (BO) for Text

Some researchers are raising attention to HPO of NLP algorithms. Wang et al. [22]
introduce a multi-stage Bayesian optimization framework for efficient HPO of NLP
applications. They also empirically study the impact of the multi-stage algorithm on
hyper-parameter tuning. Different from the previous algorithms, the new algorithm
optimizes hyper-parameter successively with increasing training data. Table 3.6 lists
the set of hyper-parameters used in SVM and boosted regression trees and Table 3.7
presents the detailed values of these hyper-parameters.

The multi-stage algorithm is an extension of the standard Bayesian optimization.
In their setting, training data $|T_{train}^1| \leq \cdots \leq |T_{train}^S|$ is increasing as time grows.
At each stage, the k best configurations passed from the previous stage are first
evaluated on the current stage's training data T_{train}^s, and then the standard Bayesian
optimization algorithm are initialized with these k settings. After running all stages,
the algorithm outputs the configuration with the highest validation accuracy from
all hyper-parameters explored by all stages. They use the proposed algorithm to

Table 3.6 List of hyper-parameters used in SVM and boosted regression trees [22]

Algorithm	Hyper-parameters
SVM	Bias, cost parameter, and regularization parameter
Boosted regression trees	Feature sampling rate, data sampling rate, learning rate, # trees, # leaves, and minimum # instance per leaf

---

Algorithm 10 Multi-stage Bayesian optimization for hyper-parameter tuning(Wang et al. [22])

Input: Loss function L,
     number of stages S,
     iterations per stage $\mathbf{Y} = \langle Y_1, \cdots, Y_S \rangle$,
     training data per stage $T_{train} = \langle T_{train}^1, \cdots, T_{train}^S \rangle$,
     validation data T_{valid},
     initialization $\lambda_{1:k}$
Output: hyper-parameter λ^*
  for $stage\ s = 1\ to\ S$ **do**
    for $i = 1\ to\ k$ **do**
      L_i =Evaluate $L(\lambda_i, T_{train}^s, T_{valid})$
    end for
    for $j = k + 1\ to\ Y_s$ **do**
      V: regression model on $\langle \lambda_i, L_i \rangle_{i=1}^{j-1}$
      $\lambda_j = \arg\max_{\lambda \in \Lambda} a(\lambda, V)$
      L_j =Evaluate $L(\lambda_j, T_{train}^s, T_{valid})$
    end for
    reset $\lambda_{1:k}$ =best k configs $\in \langle \lambda_1, \cdots, \lambda_s \rangle$ based on validation accuracy L
  end for
  return $\lambda^* = \arg\max_{\lambda_j \in \{\lambda^{Y_1}, \cdots, \lambda^{Y_S}\}} L_j$

Table 3.7 Hyper-parameters and the corresponding values used in SVM and boosted regression trees [22]

Hyper-parameters	Values
n_{min}	$\{1,2,3\}$
n_{max}	$\{n_{min}, \cdots, 3\}$
Weighting scheme	$\{$tf, tf-idf, binary$\}$
Remove stop words?	True, false
Regularization	l_1, l_2
Regularization strength	$\{10^{-5}, 10^5\}$
Convergence tolerance	$\{10^{-5}, 10^{-3}\}$

solve two tasks, including classification and question answering. For classification task they use the Yelp dataset [23]. The document includes question titles, question contexts and best answers. There are 140,000 training samples and 5000 testing samples. For question answering task, they use a commercial QA dataset containing about 3800 unique training questions and a total of 900,000 feature vectors. They tune the hyper-parameter for two ML algorithms: SVM for classification and boosted regression trees for question answering [24].

In comparison with state-of-the-art Bayesian optimization [25] and the Bayesian optimization only applied on a small subset of data for speed, the proposed method is superior across a wide range of time values and has great performance in speed. The new method can also directly help the efficiency of multiple parallel runs, as well as runs across different datasets.

Wang et al. [32] combine elements of Bayesian optimization and Simultaneous Optimistic Optimization (SOO) and introduce a new algorithm named Bayesian Multi-Scale Optimistic Optimization (BaMSOO) that eliminates the need for auxiliary optimization of the acquisition function in Bayesian optimization. They use the new method to optimize the parameters in a term extraction algorithm raised by Parameswaran et al. [33]. They compare the performance of BaMSOO, GP-UCB and SOO in automatically tuning the 4 primary free parameters of the algorithm. The objective function to is the F-score of the extracted terms. They run experiments on the GENIA corpus [34], which contains 2000 abstracts from biomedical articles. The result shows the new method outperforms standard Bayesian optimization with GPs and SOO, while being computationally efficient. The paper also provide a theoretical analysis proving that the loss of BaMSOO decreases polynomially.

3.3.2 Sequential Model-Based Optimization (SMBO) for Text

How to represent input text is an important problem in NLP, that makes a great impact on the performance. Given a particular text dataset and classification task, whether to use higher-order n-grams means an urge for more features, stronger regularization and more training iterations. Yogatama et al. [26] use a sequential model-based optimization (SMBO) technique to optimize the space of choices of machine-learned models of text. They apply the technique to logistic regression on classification tasks. The proposed method performs better than linear baselines previously reported in the literature. In some cases, their method is even competitive with more sophisticated non-linear models trained using neural networks.

The training data is $d_{train} = \langle \langle d_{i_1}, d_{o_1} \rangle, \cdots, d_{o_n} \rangle$, where each d_{i_*} is a text document and each d_{o_*} is output. The goal is to optimize a performance function f (e.g., classification accuracy, F_1 score,etc.) on the held-out dataset d_{dev} of the ML model. They consider linear classifiers with the following form:

$$c(d, i) = \text{argmax}_{o \in O}\, w_o^T x(d, i), \tag{3.9}$$

where the coefficients $w_o \in \mathbb{R}^N$ are learned using logistic regression on the training data according to output o. Let w denote the concatenation of all w_o, then the performance function f is a function of the held-out data d_{dev}, x, w, and d_{train}. The aim is to maximize f with respect to x.

Their experiments consider representational choices and hyper-parameters for text categorization problems. For SMBO, they use the HPOlib library [27]. The logistic regression trainer wraps the LIBLINEAR library [28], based on the trust region Newton method [29] and a specification of hyper-parameters. They optimize text representation based on the types of n-grams used, the type of weighting scheme, and the removal of stopwords. For n-grams, they optimize two parameters, minimum and maximum lengths. (Then, all n-gram lengths between the minimum and maximum are used.) For weighting scheme, they consider term frequency, tf-idf, and binary schemes. Last, they also choose whether they should remove stopwords before constructing feature vectors for each document.

A. Random Search and Hill Climbing for Text

One of the hot topics in NLP is how to do sentiment analysis, user analysis or text classification on a large number of text data generated in social media, blogs, online newspapers and other environments. Tellez et al. [30] propose a text classifier μTC and use random search and hill climbing to optimize the hyper-parameters of this highly parametric sentiment classifier. The search space can reach close to 4 million configurations and each configuration needs several minutes to be evaluated, so the exploration of the parameter space is expensive.

First of all, they proposed a framework to create a text classifier regardless of both the domain and the language and based only a training set of labeled examples. They find a competitive text classifier for a given task among a (possibly large) set of candidates classifiers. A text classifier is represented by the parameters that determine the classifier's functionality along with the input dataset. The search of the desired text classifier should be performed efficiently and accurately, in the sense that the final classifier should be competitive concerning the best possible classifier in the defined space of classifiers.

They use two fast meta-heuristics, random search and hill climbing algorithms. Random search selects the best performing configuration among the set C' randomly chosen from C,

$$\text{argmax}_{c \in C'} \, score(c). \tag{3.10}$$

The optimization process is denoted by a quadruple $(C, \mathcal{D}, score, opt)$, where C is the μTC space, $\mathcal{D}$ is the training set of labeled texts, $score$ is the function to be maximized, and opt is a combinatorial optimization algorithm that uses score and $\mathcal{D}$ to find an almost optimal configuration in C.

Hill climbing explore the configuration's neighborhoods N(c) of an initial setup c and then greedily update c to be the best performing configuration in N(c). The process is repeated until there is no improvement, that is,

$$score(c) \geq max_{u \in N(c)} score(u). \tag{3.11}$$

They guarantee a better performance by applying a hill climbing procedure over the best configuration found by a random search.

They consider a generic large configuration space, defined by the tuple $(\mathcal{T}, \mathcal{G}, \mathcal{H}, \Psi)$. Where $\mathcal{T} = T_i$ is the space of transformation functions, $\mathcal{G} = G_i$ is the set of tokenizer functions, $\mathcal{H}$ is a set of functions that transform a bag of tokens v into a vector $\mathbf{v}$ of dimension d, Ψ is the set of functions that create a classifier for a given labeled dataset as knowledge source. In this way, a set of all possible configurations C of the μTC space can be defined as:

$$C = T_1 \times \cdots \times T_{|\mathcal{T}|} \times G_1 \times \cdots \times G_{|\mathcal{G}|} \times \mathcal{H} \times \Psi. \tag{3.12}$$

In fact, there are 81 possible tokenizers, 32 weighting combinations. So, the configuration space contains more than 3.3 million configurations. To find the best one, they create a graph where the vertex set corresponds to C, and the edge set corresponds to the neighborhood of each vertex. In this way, the more adjacent nodes in the graph correspond to more similar configurations. This paper also uses some strategies to avoid over fitting—they use k-fold cross validation technique and binary partition to select the model.

B. Others Approaches

Mehndiratta et al. [31] identify the sarcasm in the textual data using deep learning models, i.e., convolutional neural networks (CNN) and recurrent neural networks (RNN). They measure the impact of the training data, number of epochs and amount of dropout in the network. The paper also discusses the impact of various embedding such as GloVe and fastText on the dataset when converting the same dataset into vectors via different word embeddings. They measure the influence of various parameters on the very large scale Reddit corpus:

- Dropout: Consider 15, 25, and 35% dropout in the networks. This parameter minimizes the effect of over-fitting as well as under-learning (when system performance is very poor) of the network. The larger dropout makes the network to evolve and learn more independently.
- Epochs: The number of epochs also determines whether the network is over-fitting or under-fitting. With the increase of the number of epoch, the number of iterations of weight updating in neural network increases, and the initial under-fitting state slowly enters the optimal fitting state, and finally enters the over-fitting state. Consider 2, 4, 6, and 8 epochs in the experimental setup.

The result indicats that sometimes the loss of performance is due to the configuration of the hyper-parameters and the dataset they are trying to evaluate. By tuning the hyper-parameters, the behavior of the models is improved and the large performance gain can be observed.

3.4   NAS for Multimedia Search and Recommendation

Given that neural architecture search (NAS) aims to discover the optimal neural architectures for deep neural network based models, in order to draw a clearer picture of this topic, we start by a brief introduction on the tasks of multimedia search and multimedia recommendation from the perspective of deep architectures, with some high-level frameworks and representative models presented. We further analyze the potential application pointcuts of NAS on both multimedia search and recommendation, i.e., feature extractor and modality fusion. Finally, we propose a comprehensive introduction of multimodal NAS, a cutting-edge branch of NAS for multimodal data which is of great potential value for multimedia search and recommendation.

3.4.1 Multimedia Search with Deep Architectures

From the perspective of methodology, a popular solution to cross-modal similarity search is compact coding [63], which includes hashing and quantization, converting both the query and documents into compact representation with short codes (binary codes or embedding vectors). Also, other solutions like the state-of-the-art methods in Ad-hoc Video Search (AVS) [61] are often of the same spirit. A generic framework for cross-modal similarity search with the style of compact encoding is illustrated in Fig. 3.9. Intuitively, we firstly leverage learnable encoders to map the query and documents into the same vector space and then design several similarity-related learning objectives (e.g., triplet loss based on dot product similarity) to train the encoder model.

Recently, Deep Neural Networks (DNNs) are taken as typical encoders to learn the representations of both query and documents, which achieve better multimedia search performance than conventional models. For example, in the w2vv++ model [61] for AVS, the query sentence is encoded as a vector by inputting the bag-of-words vector, word2vec embedding, and Gated Recurrent Unit (GRU)-based encoding to a Multi-Layer Perceptron (MLP) network, while the document videos are encoded by visual-features extracted by pre-trained Convolutional Neural Networks (CNNs) like ResNet-152 [58]. For another example, in semi-supervised

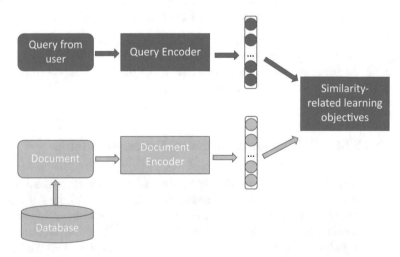

Fig. 3.9 A general framework for multimedia search

deep quantization [66], the query texts (or images) and document images (or texts) are encoded by different Deep AutoEncoder (DAE), guided by the summation of three different loss functions. In fact, these deep encoders for different media types are the pointcuts for NAS application in multimedia search, which will be detailed later.

3.4.2 Multimedia Recommendation with Deep Architectures

From the perspective of methodology, a typical deep learning model for CB-MMRS is often composed of three modules: feature extractor, feature fusion module and prediction module. A generic framework is illustrated in Fig. 3.10. Intuitively, the *feature extractor* firstly learns the representations of the items (and sometimes the user) in each modality based on the content. Then, the *feature fusion module* fuses the representations of multiple modalities to get item (and sometimes user) embeddings. Finally, the *prediction module* is fed with the embeddings of user and items and calculates the recommendation scores (often by dot product similarity) for each candidate item. Note that the main difference between the framework for CB-MMRS and ordinary-item recommendation is the feature fusion module, since the multimedia items contain multimodal features which need suitable fusion mechanism.

One can use different deep models to fill the three modules. For example, attentive collaborative filtering (ACF) model [55] takes a component-level attention module as the feature fusion module and an item-level attention module as further fusion in prediction module. Specifically, a component refers to a detected object in

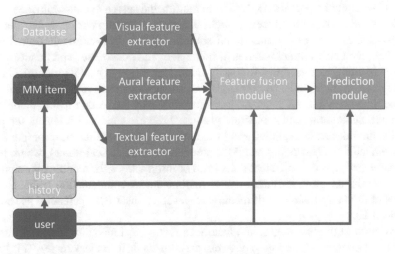

Fig. 3.10 A general framework for multimedia recommendation

an image or a frame of a video, the feature of which is extracted by pretrained CNNs. The feature fusion of different components is attended by user, thereby reflects the user preference on each component. Similarly, the item-level attention reflects the user preference on each item of his/her history. For another example, in multimodal graph convolution network (MMGCN) model [67], users and items are firstly mapped into the same vector subspaces of each modality of micro-video (i.e., video, audio and text) and the feature extractor is a stacked GCN on each subspace, where the graphs are user-item bipartite graphs. Then, a linear combination of user and item representations in different modalities is adopted as a simple feature fusion module. Both ACF and MMGCN leverage Bayesian Personalized Ranking (BPR) in prediction module to finally predict recommendation scores and train the whole framework.

As described above, literature for multimedia recommendation artificially designs all the three modules. Although extensive experiments on real-world datasets have demonstrated the effectiveness of these designs, we could still seek for automatically searched architectures to reduce human effort and prior and also achieve better recommendation performance.

3.4.3 Applying NAS on Multimedia Search and Recommendation

We have briefly introduced the concept and the most typical tasks of multimedia search and recommendation from the prospective of deep architectures, and discuss generic frameworks to solve these tasks with state-of-the-art deep learning based examples. Now let us have a closer look at the common spirit of the two tasks and how NAS could be potentially applied to multimedia search and recommendation.

In essence, both multimedia search and recommendation frameworks consist of two stages: *feature extraction* and *modality fusion*. In multimedia search (see Fig. 3.9), we firstly extract features of both query and documents, and then fuse the two embedding vectors with a similarity-related operation (e.g., dot product). On the other hand, in multimedia recommendation, we firstly extract single-modal features of multimedia items and then fuse multi-modalities by the feature fusion module to get multimedia item (and sometimes also user) representations. Moreover, the prediction module can be regarded as a further modality fusion of user embeddings and item embeddings (including candidate item and user historical items), where *user* and *item* are viewed as different modalities of informations for recommendation. In other words, the generic framework of multimedia recommendation contains two levels of modality fusion: multimedia-item-level and user-item-pair-level as shown in Fig. 3.10.

For each of the two stages, i.e., feature extraction and modality fusion, different DNNs are adopted according to different multimedia data and task target. This two-stage framework also inspires us to reduce the need for human expert architecture design and achieve automatic model design. That is, NAS approaches can also be applied to these two stages. To be more specific, we will discuss what role NAS could potentially play in each of the two stages.

A. NAS for Feature Extraction

The stage of feature extraction maps the single-modal item data into an embedding space. In multimedia search scenario, the "item data" refers to the query and documents. Traditionally, this mapping is usually supported by pretrained DNNs. For example, CNNs (e.g., ResNet) pretrained on large-scale datasets (e.g., ImageNet) are adopted to capture "concept vectors" of image or video data in multimedia search and recommendation literatures. Additionally, pretrained language models like BERT [57] are also frequently used to extract dense embeddings of sequential data of text and audio. On the other hand, other neural networks like attention [55], AutoEncoder [66], and GCN [67] are also designed to achieve more learnable feature extraction.

However, both pretrained and specially designed deep models are not guaranteed to be superior on the certain multimedia search or recommendation task because expert knowledge is required to design these models. Therefore, we could apply NAS to search for the best neural architecture in the operation space for the feature extraction of specific modality. To this end, most of state-of-the-art NAS approaches, e.g., reinforcement learning based NAS [65, 70] or differentiable NAS [62], could be leveraged to achieve high performance on downstream multimedia search or recommendation tasks. It is worth mentioning that in Sect. 3.4.4, we discuss NAS for individual modalities of text, audio, image and video, and all the methods in Sect. 3.4.4 are effective for feature extraction stage here.

B. NAS for Modality Fusion

The stage of modality fusion could happen at two levels: item-level and decision-level. To begin with, item-level modality fusion happens when the item data contains multiple media types (e.g., full micro-video = video + audio + subscripts). This level corresponds to the "feature fusion module" of multimedia recommendation framework. At this level, modality fusion stage tries to fuse the feature embeddings of each modality of a multimedia item to get overall representation of that item. Traditionally, item-level modality fusion is simply designed as concatenation or linear summation of single modal features. In some literatures this simple fusion is also followed by a Multi-Layer Perceptron (MLP) to make it more parameterized.

Moreover, decision-level modality fusion happens when the whole framework finally fuses all the item (and user) representations together and outputs the decision or prediction. This level corresponds to the operations in "similarity-related learning objectives" of multimedia search framework, and the "prediction module" of multimedia recommendation framework. At this level, multimedia search models fuse the "modalities" of query and documents, and multimedia recommendation models fuse the "modalities" of user and multimedia items. Conventionally, decision-level modality fusion takes the simple operation of dot product to calculate the similarity between query and specific document and the preference of user over specific candidate item.

It is worth mentioning that besides the fact that all the two levels of modality fusion modules mentioned above are artificially designed, these fusion strategies are all "late fusion". These characteristics leaves room for improvement with the help of NAS.

In NAS literature, a cutting-edge branch, multimodal NAS, emerges recently and raises the possibility for searching for superior neural architectures for the representation learning and fusion of multimodal data. As we detail in Sect. 3.4.4, most multimodal NAS approaches focus on how to search for a better fusion strategy [64, 69] instead of simply applying late fusion of vector concatenation. Other multimodal NAS methods also simultaneously focus on how to search for both better feature extraction and modality fusion together [54]. Therefore, multimodal NAS can be applied to both item-level and decision-level modality fusion to further boost the performance under various multimedia search and recommendation scenarios.

3.4.4 Multimodal NAS in Search and Recommendation

Multimodal NAS is a cutting-edge field that aims to apply NAS to multimodal tasks, enabling machine to automatically find the optimal model for any given multimodal task and dataset. As mentioned before, feature extraction and modality fusion are the two pointcuts for the application of NAS in multimodal tasks. To provide inspirations about multimodal NAS, we will give a brief introduction about three current representative works, namely MFAS [64], MMnas [69] and RandomNet [54], which attempts to tackle the first two problems from different perspectives. Table 3.8 gives a brief comparison of these models.

A. Multimodal Fusion Architecture Search (MFAS)

Motivation In a multimodal setting , it is common to exploit the feature extraction models and fuse in shared space outputs of the layers for subsequent downstream tasks. Fusion can be at deepest layers, a.k.a late fusion, or at shallow layers, a.k.a early fusion. As reported by Li et al. [60], the performance differs in specific tasks— it's hard to say one is absolutely better than the other. Recently, attempts of fusion across all layers have been made to get better performance in multimodal tasks [68]. However, they are all hand-crafted models whose performance depends on tasks and datasets. Therefore, as is shown in Fig. 3.11, the MFAS [64] model proposes

Table 3.8 Summary of three current papers in multimodal NAS

Model	Motivation	Search space	Search strategy	Automatic degree
MFAS [64]	Automatic modality fusion for classification	Fusion graph	SMBO	Predefined feature extractors
MMnas [69]	Automatic cross-modal feature extraction for image-text tasks	Feature extractor composition	Iterative update model weights and architecture weights	Predefined fusion graph
RandomNet [54]	Fully automatic multimodal learning	Feature extractor composition and fusion graph	Random	–

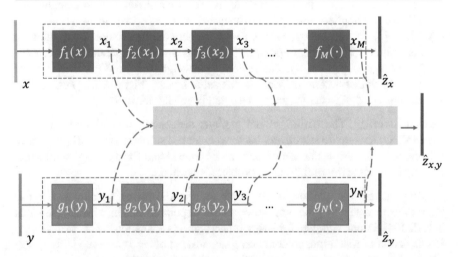

Fig. 3.11 General structure of a bi-modal fusion network. Top: a neural network with several hidden layers (grey boxes) with input $\mathbf{x}$, and output $\hat{\mathbf{z}}_{\mathbf{x}}$. Bottom: A second network with input $\mathbf{y}$, and output $\hat{\mathbf{z}}_{\mathbf{y}}$. In MFAS, finding efficient fusion schemes (yellow box and dotted lines) is the focus

to exploit NAS to automatically search the optimal fusion model for multimodal classification tasks given any datasets.

Notations The multimodal dataset is composed by pairs of input and output data $(\mathbf{x}, \mathbf{y}; \mathbf{z})$, where $\mathbf{x}$ accounts for the first modality, $\mathbf{y}$ for the second one, and $\mathbf{z}$ for the supervision labels. Two functions $\mathbf{f}(\mathbf{x})$ and $\mathbf{g}(\mathbf{y})$, are pretrained and fixed models for two modalities, which take $\mathbf{x}$ and $\mathbf{y}$ as inputs, and output $\hat{\mathbf{z}}_{\mathbf{x}}$ and $\hat{\mathbf{z}}_{\mathbf{y}}$, which are estimates of the ground-truth labels $\mathbf{z}$. Furthermore, functions $\mathbf{f}$ and $\mathbf{g}$ are composed of M and N layers, respectively, whose subfunctions are denoted by $\mathbf{f}_l$ and $\mathbf{g}_l$. For layer l, $\mathbf{x}_l = (\mathbf{f}_l \circ \mathbf{f}_{l-1} \cdots \circ \mathbf{f}_1)(\mathbf{x})$ and $\mathbf{y}_l = (\mathbf{g}_l \circ \mathbf{g}_{l-1} \cdots \circ \mathbf{g}_1)(\mathbf{y})$. Outputs of the deepest fusion layer, $\hat{\mathbf{z}}_{\mathbf{x},\mathbf{y}}$ are used for classification . Then the problem is to choose which features to fuse and how to mix them.

Search Space Each fusion layer l combines three inputs: the output of the previous fusion layer and one output from each modality. This is done according to the following equation:

$$\mathbf{h}_l = \sigma_{\gamma_l^p} \left(\mathbf{W}_l \begin{bmatrix} \mathbf{x}_{\gamma_l^m} \\ \mathbf{y}_{\gamma_l^n} \\ \mathbf{h}_{l-1} \end{bmatrix} \right), \tag{3.13}$$

where $\gamma_l = \left(\gamma_l^m, \gamma_l^n, \gamma_l^p \right)$ is a triplet of variable indices. The three elements of this triplet is the feature from the first modality, the feature from the second modality and non-linearity is applied, respectively. Also, $\gamma_l^m \in \{1, \cdots, M\}$, $\gamma_l^n \in \{1, \cdots, N\}$, and $\gamma_l^p \in \{1, \cdots, P\}$. In the case of $l = 1$, namely the first fusion layer, the term h_{l-1} is omitted. The number of possible fusion layers, a search parameter, is denoted by L, so that $l \in \{1, \cdots, L\}$. The fusion layer weight matrix $\mathbf{W}_l$ is trainable and weights are shared among matrices in the same layer l. Figure 3.12 presents two realizations of MFAS search space on a small bi-modal network.

Search Strategy The MFAS model exploits sequential model-based optimization(SMBO) strategy to search architectures, which is similar to EPNAS [64]. Since the unimodal feature extraction models are pretrained and fixed during architecture search, the complexity of the search problem is limited.

MFAS is the first one to directly tackle multimodal fusion for classification as an architecture search problem. However, as is pointed out in MMnas [69] that MFAS [64] is not suitable for tasks other than classification since MFAS only learns fusion of modalities in common space instead of respective modality space. Moreover, feature extractors are pretrained and fixed in MFAS, which makes it less flexible. To show how two problems mentioned above are tackled, we then make a brief introduction of MMnas.

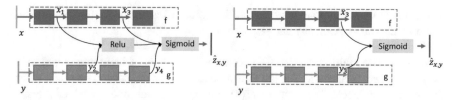

Fig. 3.12 Two realizations of MFAS search space on a small bi-modal network. Left: network defined by $\left[\left(\gamma_1^m = 1, \gamma_1^n = 2, \gamma_1^p = 1 \right), \left(\gamma_2^m = 3, \gamma_2^n = 4, \gamma_2^p = 2 \right) \right]$. Right: network defined by $\left[\left(\gamma_1^m = 3, \gamma_1^n = 3, \gamma_1^p = 2 \right) \right]$

B. Deep Multimodal Neural Architecture Search (MMnas)

Motivation Existing state-of-the-art approaches in different multimodal down-stream tasks, like image-text matching(ITM), visual Grounding (VG), visual question answering (VQA), etc., are almost task-specific and unable to be generalized to other tasks. One promising general multimodal framework is multimodal-BERT model family which is pretrained in large-scale datasets. Multimodal-BERT performs well to transfer knowledge between different tasks, but the computational costs are usually very high, which severely limits their applicability. Therefore, the MMnas model introduces NAS into multimodal learning to automatically search the optimal architecture for various tasks instead of designing a generalized framework.

Notations The multimodal dataset is composed by pairs of input and output data $(X, Y; Z)$, where X accounts for text features, Y for the image features, and Z for the supervision information. $X^{(i)}$ and $Y^{(i)}$ represent the outputs of the i-th layer respectively. M and N denote the number of model layers for two modalities. R is apriori relationship features of X and Y.

Search Space The MMnas framework is composed of two pairs, unified encoder-decoder backbone and task-specific heads. The former one learns the multimodal interactions with stacked encoder and decoder blocks, where each block is searched within a set of predefined primitive operations, including Self-Attention (SA), Guided-Attention (GA), Feed-Forward Network (FFN) and Relation Self-Attention (RSA) operations. Then the operation blocks are stacked by

$$X^{(i)} = b_{\text{enc}}^{(i)}\left(X^{(i-1)}\right), \tag{3.14}$$

$$Y^{(i)} = b_{\text{dec}}^{(i)}\left(Y^{(i-1)}, R, X^{(M)}\right). \tag{3.15}$$

Search Strategy Denote a supernet as $N(\mathcal{A}(\theta), W)$ that encodes the whole search space $\mathcal{A}$ of MMnas, where W and θ correspond to the model weights and architecture weights of all the possible operations in the supernet, respectively. The optimal architecture is obtained by minimizing the expectation with respect to θ and W jointly:

$$\left(\theta^*, W^*\right) = \underset{\theta, W}{\text{argmin}} \mathbb{E}_{a \sim \mathcal{H}(\theta)}[\mathcal{L}(N(a, W))]. \tag{3.16}$$

It adopts an iterative algorithm to optimize the architecture weights θ and the model weights W alternatively.

As is demonstrated in Fig. 3.13, MMnas is capable of automatically constructing a model for a given multimodal task, which is composed of cross-modal feature extractors and task-specific heads. For text modality, the optimal unimodal feature extractor is to search, while for image modality, the feature extractor for searching

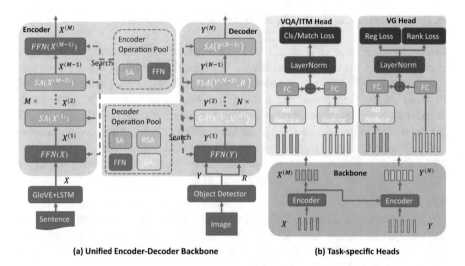

(a) Unified Encoder-Decoder Backbone   **(b) Task-specific Heads**

Fig. 3.13 The flowchart of the MMnas framework, which consists of (**a**) unified encoder-decoder backbone and (**b**) task-specific heads on top the backbone for visual question answer (VQA), image-text matching (ITM), and visual grounding (VG). Note that the searched architecture shown in (**a**) is only a schematic example

is not only extracting unimodal features, but also fusing knowledge from text modal simultaneously. However, the way of fusion is actually fixed, since the image decoder can only take inputs of prediction-level text features $X^{(M)}$ and previous image layer features $Y^{(i-1)}$. That is to say , the "fusion graph" is actually manually designed, but not automatically searched as MFAS does. The next model we will introduce, namely RandomNet, is proposed to simultaneously search both feature extractors and "fusion graph" automatically .

C. Fully Automatic Neural Architecture Design for Multimodal Learning (RandomNet)

Motivation The performance of the searched model structures may not be an adequate evaluation of a NAS method. From the perspective of the motivation of AutoML, minimization of degree of human intervention should also be one of the evaluation factors. Thus, as is illustrated in Fig. 3.14, RandomNet [54] seeks a fully automated NAS method for multimodal learning with little human expert knowledge, high flexibility and generality. To reduce the degree of human intervention, both unimodal feature extractors and model fusion network should be automatically searched in different modalities and tasks.

Search Space The search space is divided into two major components: feature extractors and fusion network.

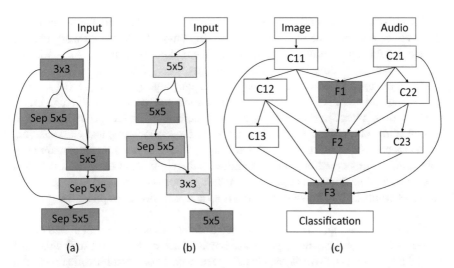

Fig. 3.14 Best cell and network structure found by RandomNet. From left to right: (**a**) image cell, (**b**) audio cell, (**c**) fusion network connectivity. 3×3 and 5×5 indicate regular convolution with the specific kernel dimension, sep indicates depthwiseseparable convolution. Colors describe the activation type: orange: identity, yellow: Tanh, green: sigmoid, pink: ReLU

RandomNet first searches cells and the final architecture is obtained by stacking these cells. Each cell is a Direct Acyclic Graph (DAG), and each node is composed by an operation O and an activation A. The operations are: 3×3 and 5×5 convolutions, 3×3 and 5×5 depthwise-separable convolutions, max-pooling and average pooling. $A = \{$ ReLU , Tanh, Identity, Sigmoid $\}$. Let L be the number of layers inside a cell, then L is fixed to 5 in the paper.

The fusion part is similar to MFAS. The difference is RandomNet allows a fusion layer at a given depth d with $d = 1, \ldots, D$ and $D \geq C$ to connect to any cell C at a lower or equal depth. More formally, a fusion layer f_d is defined by the following equation:

$$f(x_d, y_d, f_{d-1}) = \sigma_d \left(W_d \begin{bmatrix} x_d \\ y_d \\ f_{d-1} \end{bmatrix} \right), \tag{3.17}$$

where $\sigma_d \in A$, W_d indicates the trainable fusion layer weights and x_d is a vector where the probability of each element is sampled from a Bernoulli distribution with $p = 0.5$. y_d is obtained in a similar fashion. In the case of $d = 1$, namely, the first fusion layer, the term f_{d-1} is omitted.

Search Strategy Random search is adopted for simplicity and flexibility. It shows that random search is still a simple but efficient search method compared to other methods, It requires no other parameters which is of great importance for minimizing human intervention.

Recently, with the rapid development of search algorithm and spatial design to improve the performance of NAS, an increasing amount of research work focuses on the typical applications of NAS in different fields and tasks. As a method of automatically finding better neural structure for a given task, NAS has been successfully used in computer vision, natural language processing (NLP), audio processing and other fields. Specially, related computer vision tasks include image classification, image segmentation, instance detection, image generation, video classification, etc; related NLP tasks include language modeling, text classification, neural machine translation, etc.; related audio processing tasks include audio classification, audio generation, etc. In addition to the application of NAS in practical scenarios, another important research direction is more abstract and extensive, such as model compression, knowledge extraction, active learning, graph learning and so on.

Generally speaking, different tasks need different architecture design and different operations to process and capture task specified information. The main challenge for the application of NAS is the search space design for neural architectures of a given task. In this section, we introduce the current research progress on application of NAS in various modality fields and tasks. We focus on the special search space design of different tasks. This section is structured as follows. In Sect. 3.5, we introduce applications of NAS in the field of computer vision. In Sect. 3.6, we introduce applications of NAS in the field of NLP. In Sect. 3.7, we introduce applications of NAS in other tasks.

3.5 NAS for Computer Vision

Based on the fact that computer vision, as an important AI area dealing with visual signals, have close relations with the family of deep neural architectures, we will discuss the growing and exciting literature on NAS for different tasks in computer vision, including image classification, dense image prediction, image restoration, image generation etc. in this section.

3.5.1 Image Classification

As one of the most fundamental task in computer vision, image classification receives the earliest and heaviest research interests. Given an image, the goal of the task is to classify it into a specific category that best describes the image. This task is the very important because it always serves as the pretrained task for other applications in computer vision. If the pretrained model, called *backbone*, receives a large improvement, then the performance on other applications will also benefit from it.

Zoph et al. [158] first propose the concept of NAS, and provide the first solution on how to apply it to image classification problems. They define a search space of CNN in chain structure and allow skip connections to provide more complex architectures. They use RNN to construct an agent to sample architectures, and use REINFORCE to optimize the agent to output the best architecture. This kind of chained macro search space leads the early design logic in NAS research.

Zhong et al. [157] design the search space with Network Structure Code, using a 5-Dimensional vector to describe every layer inside a block. The meanings of these 5 dimensions are shown in Table 3.9. In this way, they can build more flexible and powerful architectures. After building a block using the Network Structure Code, they generate the whole network in a predefined way (see Fig. 3.15), where the searched block is repeated several times combined with predetermined layers to form a larger architecture. Because the whole network is automatically generated by specifying the overall architecture, the search target is just a repeated motif. This kind of search space is called *micro search space*, in contrast to the macro search space which aims to directly find the whole architecture. The micro search space is typically smaller than the macro one, which will largely reduce the search cost and increase the transfer ability among models and tasks at different skills. Similar architecture encode design can also be found in [105].

At the same time, Zoph et al. [159] also design their search space in a micro way. The only difference is that they adopt two repeatable cells—*normal convolution cell* and *reduction convolution cell* to design their whole network. Normal convolution cell outputs the same dimension feature map as the input, and reduction convolution cell reduces the width and height by a factor of 2. The reduction cell is designed to replace the pooling layer in [157] to enable more flexible network design. This design of micro space with two kinds of cells leads the research trends of CNN architecture search.

Different from [159], Farabet et al. [120] propose to describe the search space in a hierarchical manner (Fig. 3.16). The searched model is decomposed to several hierarchies. Each level of the hierarchy is a set of acylic computation graphs and each node represents a feature map. The highest level only contains a single motif corresponding to the full architecture, while the bottom level is a set of primitive operations. The graph motifs of the lower level serve as graphic edge transform

Table 3.9 Meaning of 5 dimensions in a description vector

Dimension	Meaning
Index	The position of the current layer in a block
Type	Seven types of commonly used layers such as convolution, max pooling etc.
Kernel size	Three kinds of kernel sizes available for convolution layer and two sizes for pooling layer
Pred1	Predecessor parameters which is used to represent the index of layers predecessor
Pred2	Same as Pred1

Fig. 3.15 Auto-generated
networks on CIFAR-10(left)
and ImageNet(right). The
searched blocks are repeated
in a predefined structure

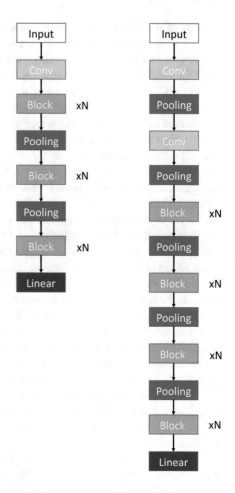

operations in higher level. They make use of evolutionary algorithm to search for
the model with best performance.

Brock et al. [77] propose a memory-bank design of neural architecture. First of
all, they design a macro framework of the neural architecture to search four blocks
at different spatial resolutions. Each block has several memory banks. Each layer
inside the block in the network simply reads a subset of memory, modifies the
data and writes the result to another subset of memory. The design of this search
space makes it possible to sample complex and branching topologies and encode
candidates into binary vectors effectively. To improve the speed, they build a model
named HyperNet to generate the weights of sampled neural architectures, and use
random search to find the models with best performance on validation set.

Following [159], Pham et al. [133] propose to describe the convolutional cell in
a Directed Acylic Graph (DAG) manner, which enables them to use weight sharing
to speed up the search process by a large scale. The models in their search space
are all sampled from the same full graph. Therefore, the parameters of an operation

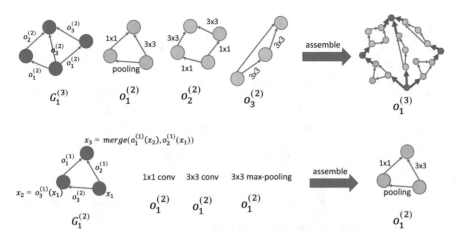

Fig. 3.16 An example of a three-level hierarchical architecture representation. The bottom row shows how level-1 primitive operations $o_1^{(1)}, o_2^{(1)}, o_3^{(1)}$ are assembled into a level-2 motif $o_1^{(2)}$. The top row shows how level-2 motifs $o_1^{(2)}, o_2^{(2)}, o_3^{(2)}$ are assembled into a level-3 motif $o_1^{(3)}$

can be shared among all the model candidates which have the same operation on the same edge. They adapt a RNN agent to sample models from the full graph, and use weight sharing to update the parameters in the full graph. The time and resources in the search procedure is therefore saved by 1000x compared to [159]. Because of the easy implementation and efficient search procedure, this design of search space and usage of weight sharing lead some research [121, 126, 146, 150].

In order to change the connection topology of architectures during the transforming operation, Cai et al. [79] describe their search space in a tree structured manner (Fig. 3.17). The operation allows replacing a single layer with multi-branch motif, which can be encoded as a tree. The input of the tree is processed and split by nodes from root to leaf, and then merged back from leaf to root to get output. The design of the tree structure search space makes it easy to transfer parameters between new models and old models, even when the topology changes greatly. They also build a tree-structured LSTM agent to extract the node feature in the current model tree and sample next tree candidate.

Wu et al. [148] also propose a micro search space, where the overall architecture is predetermined with several blocks connected in a chain manner. Some hyperparameters of these blocks are also fixed, including filter size, number of layers and stride. However, different from [159] and [157], in order to find architectures with low latency, they don't force different blocks to have the same structure. Instead, they search architectures of all these blocks together. Different from [159], the topology of each block is designed to improve efficiency, all of which are in chain structure with some skip connections. The only search target is the operations inside each block. They use Gumbel-softmax to enable back propagation and use gradient

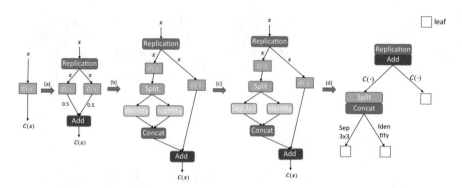

Fig. 3.17 An illustration of transforming a single layer to a tree-structured motif via path-level transformation operations, where they apply Net2DeeperNet operation to replace an identity mapping with a 3×3 depthwise-separable convolution in (c)

descent to update the architecture parameters. Delay is also considered in the design of loss function.

Similarly, Tan et al. [140] propose MnasNet with a micro search space to solve image classification problems. The difference is the hyper-parameters such as filter size, number of layers and squeeze-and-excitation ratio are also searched instead of fixed. The set of primitive operations are also designed according to MobileNetV2 [135]. MnasNet also takes latency into consideration, and use reinforcement learning to search for the best configurations.

Cui et al. [89] further relax the micro search space in MnasNet. By describing each block in a DAG style, the topology becomes more complicated and the search space is greatly expanded. To efficiently search for the best models in the huge space, they propose a coordinate search algorithm to search for architecture of blocks separately and iteratively. Combined with weight sharing, the sample efficiency is reduced to an acceptable range. However, different from DAG design in ENAS, the topology of each block is fixed and will not change during the search procedure. Only the best operations are searched on each edge, which is different from DAG in ENAS that the topology is also searchable.

Based on the typical blockwise NAS paradigm, Mei et al. [124] propose Atomnas to design the search space in a more fine-grained manner. They find that the current primitive operations in NAS for image classification have the same formula as:

$$Y = (f_1^{c'',c'} \cdot g \cdot f_0^{c',c})(X), \tag{3.18}$$

where $f_1^{c'',c'}$ and $f_0^{c',c}$ are two convolution atom operators, g stands for a channel-wise atom operator. They find that the formula above can be regarded as an ensemble model of many one-channel operations:

$$Y = \sum_{i=1}^{c'} (f_1^{c'',1}[i,:] \cdot g[i,:] \cdot f_0^{1,c}[:,i])(X). \tag{3.19}$$

They further relax the constraints that every channel operation should have the same atom operation. Then, a more fine-grained model and search space is designed:

$$Y = \sum_{i=1}^{c'} (f_{1i}^{c'',1} \cdot g_i \cdot f_{0i}^{1,c})(X). \tag{3.20}$$

This kind of search space is able to contain more powerful operators such as Mixconv in Mixnet. After design a fine-grained search space, they use DARTS to search for the best models. They also add L1 penalty of importance vector, combined with proposed pruning techniques to effectively remove dead operations during search.

In order to search the number of layers and network widths, Fang et al. [93] propose a densely connected search space. They separate the search space of NAS into three levels, specifically, from the bottom to the top as basic layer, routing block and dense super network. The basic layer transforms an input to an output with several optional primitive operators. The routing block is used to combine the outputs of several preceding routing blocks with different shapes. The densely connected super network is composed of multiple routing blocks. The number of layers can be searched by setting primitive operations in basic layers to skip connect, and the network widths can be searched by choosing blocks with appropriate widths. They then relax their search space in a continuous form and use DARTS to perform the search.

Recently, Li et al. [111] take non-local operators into consideration. Based on Mnasnet's search space, they also add non-local operator to their search space. Together with the conventional NAS procedure, it searches the positions where to put the non-local operators and the ratios of down-sampling. To achieve the search objective, they design some differentiable search algorithms to take location and down-sampling ratio of non-local operator into consideration. Then, Single-Path NAS [104] is used as search strategy to solve the whole problems.

In order to search for channel numbers and input resolutions in an affordable way, Wan et al. propose FBNetV2 [142]. For channel search, they use weight sharing and channel masking to uniform the output shapes of different channel choices and reduce memory and time cost. For input resolution search, they design a uniform zero-padding and convolve directly on the subsampled input to handle pixel and receptive field misalignment problems. For other searchable parts such as layer number, operation type, expansion ratio, they adapt a layer-wise micro search space similar to FBNet [148]. By leveraging Gumbel-Softmax to relax the hyper-parameter space, they perform the search in a differentiable manner.

3.5.2   Dense Image Prediction

Dense image prediction is a high level task in computer vision, which needs to predict every pixel with a label. This task includes semantic image segmentation, object detection, instance detection and so on. It is not suitable to transplant NAS research of image classification directly to dense image prediction, because dense image prediction needs to operate on high-resolution images, and the original operation is very different. Tremendous hand-crafted networks and operations are designed to solve these kinds of tasks, such as spatial pyramid structures [94], atrous convolutions [85], FBNet [148] and so on. Recently, there are also many researchers paying attention to perform NAS on dense image prediction tasks to let computer find the best architectures automatically.

A straightforward way of applying NAS to dense image prediction problems is using the networks searched on image classification to replace the backbone used in previous researches. Benefiting from the performance improvement from image classification, this simple method can also lead to performance gain in dense image prediction tasks. Some researchers [88, 93, 111, 124] use the searched network pretrained on Imagenet or COCO as backbone to solve the semantic segmentation PASCAL VOC 2012 task. Cui et al. [89] use the searched network from CIFAR-10 and replace the encoder of PSPNet [156] with the searched one to solve the semantic segmentation ADE20K task. Xu et al. [150] plug the searched network into object detection framework named Single-Shot Detectors (SSD) [122] to solve the object detection MS-COCO task.

There still exists a large amount of works on how to directly search the backbone (encoder) and/or decoder. These kinds of researches directly apply NAS to the object of dense image prediction tasks to further boost the performance of NAS.

A. Semantic Segmentation in Images

Liang et al. [84] present the first step towards applying the NAS directly to dense image prediction. n order to solve the task of scene annotation, they mainly study how to build an extensible and processable search space, and how to design a proxy task which is predictive and fast experimental for large-scale tasks. In their work, the feature extraction backbone is fixed, and the decoder is searched using NAS. For the space design, a recursive search space is constructed to encode multi-scale context information, which is termed Dense Prediction Cell (DPC) (Fig. 3.18). The cell is represented by a DAG consisting of several branches, with each branch using one operator to process an input to derive another output tensor. The input is selected from the last layer of network backbone feature maps and outputs from previous branch. The operations are selected from a predefined operation set, including convolution with a 1×1 kernel, 3×3 atrous separable convolution with rate $r_h \times r_w$, where r_h and $r_w \in 1, 3, 6, 9, \ldots, 21$ and average spatial pyramid pooling with grid size $g_h \times g_w$, where g_h and $g_w \in 1, 2, 4, 8$. The final output is a concatenation of the

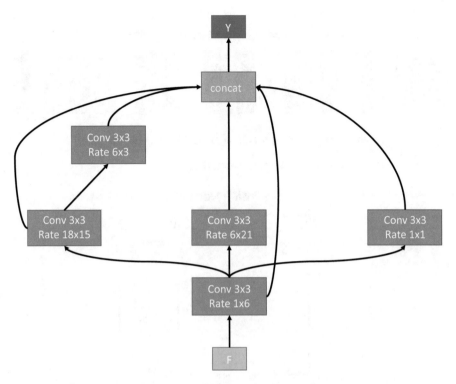

Fig. 3.18 Schematic diagram of top ranked DPC

outputs from all of the branches. The designed search space contains the previously hand-crafted state-of-the-art networks such as [86, 156]. For the proxy tasks, they select a smaller backbone and fix the parameters of backbone to save the training time. Early stopping is also adopted to further reduce the evaluation time. After several best architectures are found at the proxy task, they rerank them according to the performance after retraining using conventional experiment settings. They use random search from [100] to perform the NAS.

Li et al. [112] also design a search space for segmentation network. The overall marco-architecture of the proposed search space is shown in Fig. 3.19. The feature maps of last three stages from the backbone are used as the input of decoder. Every feature map from each stage is processed by a channel controller to change its channel number. Then, two fusion nodes (Fig. 3.20) are designed to fuse the available feature maps and derive the score map. The score map is then up-sampled by 8 times to produce the final per-pixel semantic segmentation prediction.The hyper-parameters here are the channel number of the output of the three channel controller. They use the proposed *Partial Order Pruning* algorithm to search for the decoder and take the inference time into consideration, together with the backbone searched on Imagenet using the same method to form the final semantic segmentation networks.

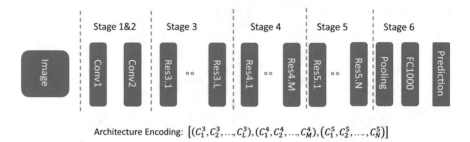

Architecture Encoding: $\left[\left(C_1^3, C_2^3, \ldots, C_L^3\right), \left(C_1^4, C_2^4, \ldots, C_M^4\right), \left(C_1^5, C_2^5, \ldots, C_N^5\right)\right]$

Fig. 3.19 A general network architecture in the search space

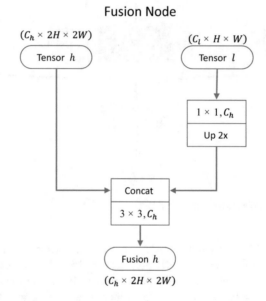

Fig. 3.20 Detailed structure of the fusion node

In the meantime, Nekrasov et al. [128] also apply NAS to semantic segmentation. Their aim is to find an efficient search framework instead of using almost 400 GPUs over 7 days like [84]. The search space is designed as shown in Fig. 3.21. Like [84], they only search for the decoder. The outputs of different groups of layers in backbone, after being processed by a 1×1 conv operation, are available for decoder. The macro architecture (on the left of Fig. 3.21) contains N more blocks, with each block takes element-wise-sum two inputs, selected from the previous blocks, after processed by a cell, whose topology is shown on the right of Fig. 3.21. Then, the output of the two blocks are concatenated and processed by another 1×1 conv operation to produce the prediction. The micro cell search space consists of M edges, with each edge selecting one output from previous edges or original input as its input. The input is then processed by one selected operation. At last, all

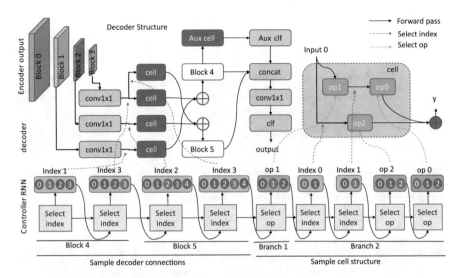

Fig. 3.21 Example of the encoder-decoder auxiliary search layout

the loose ends of the edges are summed up to be the output of this cell. They borrow reinforcement learning to search for the best architectures (the bottom of Fig. 3.21). To speed up training, they use two-stage training method with early stopping, and use knowledge distillation and intermediate supervision by auxiliary cells (see Fig. 3.21) for fast and stable converge.

Instead of searching for the small decoder like [84], Liu et al. [118] aim to directly search for the backbone for semantic segmentation tasks while keep the decoder fixed as a ASPP module, and will be then bilinear upsampled to the original resolution before producing the prediction. Following the common space design pattern in [140, 159], they propose a two-level hierarchical search space. The inner cell level is borrowed from [119, 121, 159], where each cell consists of several blocks. Each block takes the outputs of the selected two previous blocks as inputs. Two primitive operators are selected with replacement to process the given inputs. Then, the two processed feature maps are element-wise-added to produce the output of current block. The outer network level is designed specially for semantic segmentation tasks. Based on the observation of current state-of-the-art backbones, they design a multi-level macro space (see Fig. 3.22). There are four spatial resolution levels in the space. The first two layers are fixed to reduce the spatial resolution by a factor of 2. Then, the following layer connections are searched with a constraint that the resolution of next layer is either twice as large, or twice as small, or remains the same as the current layer. After relaxation of the hyper parameter space in each level, they use DARTS to search for the best architectures.

Different from previous work, Zhang et al. [155] aim to search for both backbone and decoder for semantic segmentation, and also take the model cost into consideration. The proposed search space is shown in Fig. 3.23. Two cell

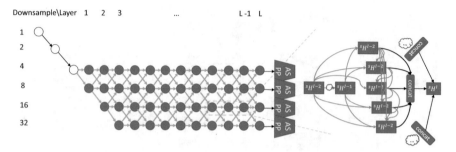

Fig. 3.22 left: the network level search space with $L = 12$. Gray nodes represent the fixed "stem" layers, and a path along the blue nodes represents a candidate network level architecture. Right: during the search, each cell is a densely connected structure

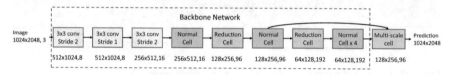

Fig. 3.23 An overview of the network structure for semantic segmentation in [155]. Each cell employs the previous two cells' outputs as its inputs

types are used for searching the backbone like the conventional practice in image classification [159]. A multi-scale cell is designed as the decoder. It takes the last two spatial resolution outputs of the backbone as inputs, and process them with 9 nodes to produce the output. Each node selects two outputs from previous nodes and original inputs, and process them using selected two operations. Then the feature map is summed up to produce the output of current node. They relax the hyper-parameter space in a continuous manner, and use gradient descent to directly update the hyper-parameter selections.

In order to investigate whether the weight sharing is effective, Bender et al. [76] conduct experiments on semantic segmentation backbone using NAS. They design a MobileNetV3-like macro architecture, and use methods similar to image classification [133] to design the whole search space. Cells are stacked into multiple circles to build the whole backbone and each cell is described using DAG. They use reinforcement learning to search for the best models and this algorithm can be generalized to search for semantic segmentation tasks.

Lin et al. [116] propose a new search space for semantic segmentation backbone search. The backbone is split into several cells and each of them is represented as a DAG (see Fig. 3.24). The node in DAG takes the outputs of all previous nodes as inputs and sum them together. The operations on the edges are to be searched. They further relax the search space using Gumbel-Softmax. For the search algorithm, they build a GCN-Guided Module, which can update the architecture weights in current cell by taking weights of previous and current cells as input. The whole search process is built in a differentiable way.

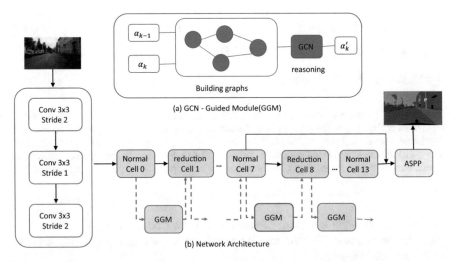

(a) GCN - Guided Module(GGM)

(b) Network Architecture

Fig. 3.24 Illustration of the Graph-Guided Network Architecture Search [116]. In reduction cells, all the operations adjacent to the input nodes are of stride two. (**a**) The backbone network, it's stacked by a series of independent cells. (**b**) The GCN-Guided Module (GGM), it propagates information between adjacent cells. α_k and α_{k-1} represent the architecture parameters for cell k and cell $k-1$, respectively, and α'_k is the updated architecture parameters by GGM for cell k. The dotted lines indicate GGM is only utilized in the search progress

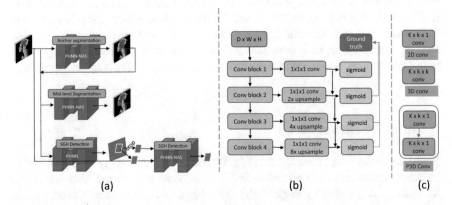

Fig. 3.25 The framework proposed in [102]. (**a**) SOARS stratifies OAR segmentation across two dimensions: distinct processing frameworks and distinct architectures where the later is executed using differentiable NAS. (**b**) Depicts illustrates the backbone network (P-HNN) with NAS, which allows for an automatic selection across 2D, 3D, P3D convolutions. (**c**) Demonstrates the NAS search space setting

Guo et al. [102] also apply NAS to organ at risk segmentation for head and neck cancer task. They first design a three-level segmentation network using P-HNN as backbone (see Fig. 3.25). Then, NAS is used to search for the best convolution block for each P-HNN backbone used. The overall connection topology is fixed and the

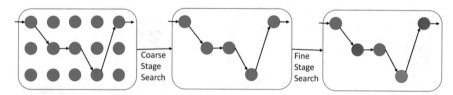

Fig. 3.26 An illustration of the proposed C2FNAS. Each path from the left-most node to the right-most node is a candidate architecture. Each color represents one category of operations, e.g., depthwise conv, dilated conv, or 2D/3D/P3D conv which are more common in medical image area. The dotted line indicates skip-connections from encoder to decoder. The macro-level topology is determined by coarse stage search, while the micro-level operations are further selected in fine stage search

only searchable target is the convolution block, where the *2D conv, 3D conv* or *P3D conv* can be selected. The differentiable architecture search algorithm is used to search for the best model.

Yu et al. [152] apply NAS on 3D medical image classification tasks. They develop a coarse-to-fine search algorithm to find the best architecture. At the coarse stage, a macro search space similar to the one in [118] is designed (see Fig. 3.26), and the target is to find the best connection topology. The model consists of several cells, with each cell can be connected to the next cell with twice of, the same, or half of resolution input. All the operations on edges are fixed to standard 3D convolution. Evolution algorithm is adopted to search for the best connection at coarse level, and each model trained from scratch. At the fine level, the connection topology is fixed, and the operations on each edge is searched. Single-path one-shot NAS algorithm with uniform sampling is used for the second stage to save computation and memory cost.

B. Object Detection in Images

Based on RetinaNet framework [117], Ghiasi et al. [99] apply NAS on object detection problems. Their aim is to find a better Feature Pyramid Network (FPN) for RetinaNet. Inspired by Zoph et al. [159], they also build their FPN search space in a recursive manner. The searched FPN cell produces the same shapes of outputs as its inputs, thus the cell can be stacked several times to produce a stronger FPN. The first cell takes M (in this work they take 5) different scales of input features, which are the last layers from every group of feature layers, from the backbone of RetinaNet as inputs and output the same number and shape of feature maps. The cell afterwards takes the outputs of previous FPN cell as inputs. Each FPN cell consists of N merging cells, with each merging cell has 4 hyper-parameters to be determined: the two input feature maps, the output resolution and the operation that merges the given two maps. The operation can be *sum* or *global pooling*, both are parameter-free. After the merge, the feature map is processed by a ReLU, a 3×3 convolution and a batch normalization layer. The last M merging cells are designed as outputs.

All the loss ends in the intermediate nodes will be added to the outputs with the same resolution. Similar to [158], they use a controller RNN to determine all the hyper-parameters in merging cells, and use REINFORCE to update the parameters of the controller. They also supervise all of the derived feature maps at different resolution with the object detection labels to enable *anytime detection*. Similar to [84], they use early stopping and small backbone to build a fast proxy tasks for searching.

Following [99], Xu et al. [149] also apply NAS on object detection tasks. They further separate the decoder of object detect network into two parts: the neck fusion part and the head network. Then, they design new search spaces for each of them (see Fig. 3.27). For the auto-fusion search space, they borrow the repeated design in [99], but allow each layer to have different topology. The search space is fully connected: each input and output stage (a certain spatial resolution feature map) has an edge that defines an operation from the primitive operation pool. For the auto-head search space, a split-transform-merge paradigm of design is used. The macro space is composed of several cells, with each cell has 7 nodes. The beginning two nodes are the outputs of previous 2 cells. The intermediate node will first select one node as input, and then process it by a selected operation from primitive operation pool. Similar to the design principal of auto-fusion space, different cell in auto-head can also have different topology. Similar to [155], they relax the hyper-parameter space to allow gradient search, and also take the computation cost of model into consideration when optimizing.

In the meantime, Wang et al. [144] also aim at searching for both the FPN and head network (see Fig. 3.28). They propose a DAG search space for FPN, where each node selects two outputs from previous nodes. They apply two independent operations on them and then merge them with addition or concatenation. The last three nodes are set to be the output of FPN network, with all the intermediate

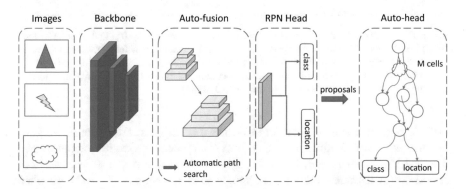

Fig. 3.27 An overview of the architecture search framework for detection, which can be stacked on any backbone and focuses on searching better architectures of the neck and the RCNN head of a detection system. The proposed Auto-fusion module aims at finding a better architecture for utilizing information from all feature hierarchy and conducting feature fusion for better prediction. The Auto-head module is constructed by different auto-searched cells in order to perform a better classification and bounding-box (bbox) prediction

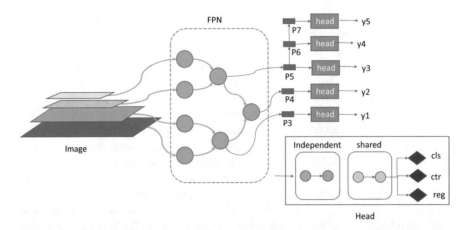

Fig. 3.28 A conceptual example of the NAS-FCOS decoder. It consists of two sub networks, an FPN f and a set of prediction heads h which have shared structures. One notable difference with other FPN-based one-stage detectors is that the heads have partially shared weights. Only the last several layers of the predictions heads are tied by their weights. The number of layers to share is decided automatically by the search algorithm. Note that both FPN and head are in the actual search space, and have more layers than shown in this figure which is for the purpose of illustration only

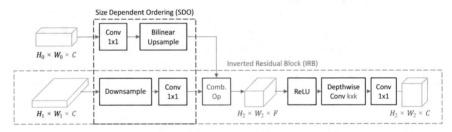

Fig. 3.29 A searchable MnasFPN block. MnasFPN re-introduces the Inverted Residual Block (IRB) into the NAS-FPN head. Any path connecting an input and a new feature, as highlighted in blue dashed rectangle, resembles an IRB. MnasFPN also employs size dependent ordering (SDO) shown in black rectangle to re-order the resizing operation and the 1×1 convolution prior to feature merging. The searchable components are highlighted in red

nodes's loose ends added to each of them. The three outputs at different resolutions, together with two additional inputs derived from processing the last output by two 3×3 stride-2 convolution, forms the inputs of head network. The head network is designed in a sequential form. The weight sharing position can also be searched after the weight of the head is shared in different resolutions. They used RL and LSTM controllers to search for the best decoder.

Following [99], Chen et al. [82] propose the MnasFPN to further improve the performance of NAS on object detection, and make the first step towards directly optimizing network architecture for mobile deployment. The overall search space is borrowed from [99]. Figure 3.29 shows a searchable block of their

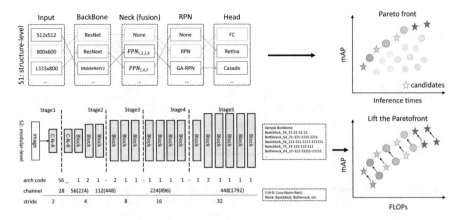

Fig. 3.30 An overview of SM-NAS for detection pipeline. A two-stage coarse-to-fine searching strategy directly on detection dataset is proposed: S1: structural-level searching stage first aims to finding an efficient combination of different modules; S2: modular-level search stage then evolves each specific module and pushes forward to a faster task-specific network

search space. Different from [99], they allow the intermediate feature map to have different channel numbers as inputs, where both the input channel number and the intermediate channel number is searchable. Instead of adding all the loose ends of intermediate nodes to the output nodes, they discard all the loose ends. This enables block count search and controls the computational costs. They also add cell-wide residuals that add residual connections between every pair of input and output features at the same resolution. The side dependent ordering is also proposed to reduce the computation cost: the 1×1 conv operation will only be applied before upsampling or after downsampling. The search framework is borrowed from Mnasnet [140], where a controller is used to propose new networks and updated using reinforcement learning.

Guo et al. [103] propose to search both the backbone and the decoder at the same time for object detection tasks and take computational cost into consideration. They split the object detection network into four peaces: backbone, neck fusion, RPN and head. They fix the PRN architecture during searching, and design the overall connection topology of the left three parts. All the left edge operations need to be searched. They associate each operation with a score on each edge to determine its importance, and relax the whole search space in a continuous way. To reduce the computational cost, they perform a two-stage search algorithm. At the first stage, the primitive operation pool of different parts will be screened according to the importance score until the whole search space is reduced to an affordable range. At the second stage, the sub primitive operation pools are fixed, and the importance score and the weight of the supernet is updated using gradient descent. The final architecture is derived by taking the operations with the highest score on every edge.

Yao et al. [151] also search the whole architecture for object detection tasks using a two stage search algorithm (see Fig. 3.30). They design a coarse-to-modular

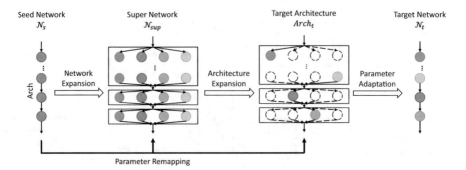

Fig. 3.31 The framework of the FNA. Firstly, the authors select an artificially designed network as the seed network N_s and expand N_s to a super network N_{sup} which is the representation of the search space. Then parameters of N_s are remapped to N_{sup}. They utilize the NAS method to start the architecture adaptation with the super network and obtain the target architecture $Arch_t$. Before parameter adaptation, the parameters of N_s to $Arch_t$ are remapped. Finally, they adapt the parameters of $Arch_t$ to get the target network N_t

two stage search algorithm to search for the Pareto front. The overall network architecture is split into four parts. The difference is that the RPN part is also searchable instead of fixed. In the coarse stage, the module inside each part of network is searched. There are several choices inside each part together with the several choices of input resolution, and the aim is to find the best combinations of modules. After the module choice is determined, the fine-grained structure of the whole network is then searched to further lift the Pareto front, which is done in the modular stage. The search space in the second stage consists only of the channel numbers and layer numbers of backbone, together with the channel numbers in FPN. They use the evolutionary search algorithm and use Partial Order Pruning method [112] to reduce the search space.

Fang et al. [92] propose a neural network adaptation method to quickly adapt the pre-trained network's parameters and architectures to the given tasks. The overall search procedure is shown in Fig. 3.31. Firstly, they expand the pre-trained network to a big supernet with each layer can have other different operation choices. The parameters in original pre-trained network is also remapped to the operations of the whole search space. Then, common single-path differentiable NAS framework like the one in [80] is adopted to search for the best architecture. After the architecture is developed, the parameter of original pre-trained network is again remapped to it. The network is then fine-tuned on the given tasks.

Jiang et al. [106] firstly introduce a novel ElixirNet to medical lesion detection. The core part of ElixirNet is Auto-lesion Block which is found by differentiable NAS. They design a novel search space according to the characteristics of medical images. They add a non-local operator to the search space, in order to encode sematic relation between region proposals which is relevant to the object detection to extract the lesion out of similar no-lesion region. The relation is formulated as a region-to-region undirected graph where node corresponds to a region proposal and edge encodes relationship between two nodes.

3.5.3 Image Restoration

Image restoration aims to restore the original image that is corrupted. It includes image inpainting and image denoising. Currently, state-of-the-art models for this task leverage generative training or Convolutional AutoEncoders (CAEs). As NAS becomes more and more popular in other research areas of computer vision, many researchers also study on how to apply NAS on image restoration problems.

Suganuma et al. [139] make the first step towards NAS application on image restoration tasks. They propose to search for the best CAEs to solve the given tasks. They design a symmetric search space for CAEs, and only focus on searching for encoders (decoders are contructed in a symmetric way). The whole network consists of several cells, with each cell first selects one previous cell as input, and applies a selected operation on it to derive the output. The output cell only selects one intermediate cell with no select operation. They use evolution algorithm as the search algorithm to find the best architectures.

Different from [139] that search for a symmetric network, Zhang et al. [153] propose to search a computation cell as the basic block and then build the final architecture by stacking the found block with different widths. The search space of inner cell is the same as [133], which is represented by a DAG. Every edge represents a transform operation, and each node will only select two inputs to sum up as its output. In order to also search for the width of each cell, they construct a supernet, each layer of which has three different width choice. The cell in current layer can only be connected to the cells in last layer with the same, twice or half of widths. They use methods in [121] to search for the best architecture, accelerating and stabilizing the search by proposing cell sharing and early stop.

3.5.4 Image Generation

Image generation is an important and tough task that aims to generate natural images condition on noises or given images. Generative Adversarial Networks (GAN) are often used for image generation problem. Applying NAS on GAN is a harder task compared to other tasks because GAN is unstable when training and easy to collapse even it is hand-crafted. The evaluation of GAN is also not as straightforward as other tasks [101]. Recently, many efforts have been devoted to apply NAS on GAN to further boost the performance of related tasks.

Gong et al. [101] propose the first NAS framework for GANs, which is called AutoGAN. To stabilize the GAN searching and training procedure, they only focus on searching for generator of GANs, and progressively grow the search space of

GANs by adding more layers. The discriminator will also grow in a predefined way as the search space is enlarged. The GAN architecture is represented as a network that consists of several cells, with each cell can be represented by a $(s + 5)$ element tuple $(skip_1, \ldots, skip_s, C, N, U, SC)$, where the beginning s elements stand for the skip connection from previous cells, C stands for type of convolutions, N stands for normalization types, U stands for up-sampling methods and SC stands for in-cell shortcut. They use dynamic-resetting to avoid GANs being collapsed affecting the shared weights. Like [158], an RNN controller is adopted to sample GANs network, whose inception score on validation dataset is rewarded to controller to update its weights.

Following [101], Gao et al. [97] propose to search for both the generators and discriminators in a differentiable NAS framework. They define the search space of GANs by stacking up-cells and down-cells, which will upsample and downsample the resolutions respectively. The generator is composed of 3 Up-Cells and the descriminator has 4 down cells. Each cell is a pre-defined DAG, with the operations on edge to be searched (See Fig. 3.32). The up-sample and down-sample operations are only selectable by edges from the inputs and to the outputs respectively. Like [121], they also make the search space to be continuous using softmax, and propose an adversarial searching algorithm. They directly fix one part of GANs as supervise signal to update the architecture and parameters of another. In this way, the time-consuming supervise signal like inception score in [101] can be avoided.

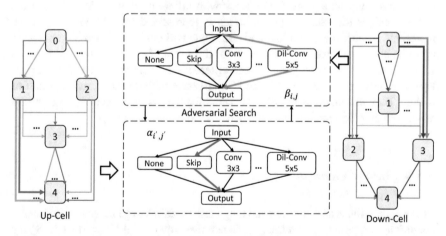

Fig. 3.32 The search space of Up-Cell and Down-Cell. The architectures of both Up-Cell and Down-Cell will continuously promote each other in an adversarial manner

3.5.5 Other Applications

In addition to the research areas mentioned above, some other computer vision tasks can be improved by NAS. Due to the relatively small degree of attention, we list it in this chapter.

Ryoo et al. [134] propose the first NAS framework towards solving video classification architectures. The Assemblenet (see Fig. 3.33) is a DAG with each node as a sub-network block and each edge represents the information flow. The topology of each node is designed following ResNet, with m (a fixed number) of interleaved 2D and (2+1)D residual modules. The whole Assemblenet consists of four level of nodes, with each level has different m. The edge is constraint to only begin at lower level nodes and end at higher level nodes. They use evolutionary algorithm to search for the best architectures. A family of randomly connected architectures are initialized and is mutated by changing block connectivity based on the learned weights, modification of temporal resolutions of a convolutional block or merge or split a block.

Li et al. [114] propose a NAS algorithm formulated by Gussian process, and firstly apply it on face recognition tasks. They adopt MobileFaceNet [87] with a width multipliers of 0.75. The search space contains the expansion rate $n \in \{2, 4, 6\}$ and whether the squeezing and excitation mechanism is enabled or not

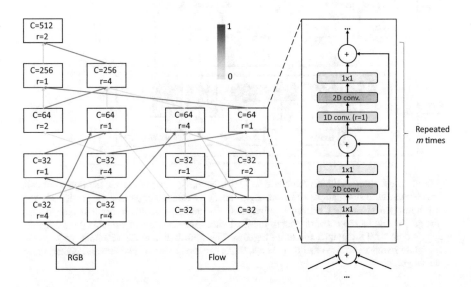

Fig. 3.33 AssembleNet with multiple intermediate streams. Darker colors of connections indicate stronger connections. At each convolutional block, multiple 2D and (2+1)D residual modules are repeated alternatingly. The network has 4 block levels (+ the stem level connected to raw data). Each convolutional block has its own output channel size (i.e., the number of filters) C and the temporal resolution r controlling the 1D temporal convolutional layers in it

for each block. Using the Gussian process based NAS algorithm, a model with less parameters and FLOPs but higher or competitive accuracy are derived.

Peng et al. [130] firstly apply NAS on Skeleton-based human action recognition tasks using GCN. The search space is constructed as the one in Fig. 3.34. They design 8 function modules for generating one computation cells, and they stack several cells to construct the final architecture. The 8 function modules contain one spatial module, one temporal module, one spatial-temporal module and five modules from each step of Chebyshev K-hop connections. Instead of fixing the cell structures throughout the whole network like [121, 133, 150], they allow different function modules selected for different layers. The method is adopted for quick and efficient search in the defined search space.

Li et al. [110] apply their search algorithm on searching GCN architectures to solve 3D object classification tasks. They design the search space of GCN similar to the one in [121]. The whole network consists of several stacked cells, and each cell is a DAG with multiple nodes. The previous two cells outputs are used as the first two nodes of current cell. The only difference is that the primitive operation pool only consists of GCN-specific operations. The Sequential Greedy Architecture Search (SGAS) is used to find the best models.

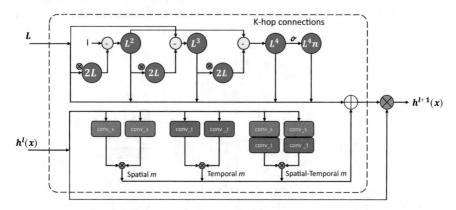

Fig. 3.34 Illustration of the search space in [130]. Here, $\otimes$ denotes matrix multiplication. $\oplus$ is the element-wise summation. There are eight function modules for generating graphs. The top part is a implementation of Chebyshev polynomial. The authors also add its separate components to the graphs and let the network choose the final ones. The bottom part contains three dynamic graph modules. All the graphs are added together. The contribution of each module works as the architecture parameters. Note that there is a softmax function before the summation operation for dynamic graphs

3.6   NAS for Natural Language Processing

Given that natural language processing, as an important area that focuses on analyzing linguistics, also benefits in the development of deep learning models, we will discuss research works on NAS for language modeling, text classification as well as other tasks in the area of natural language processing in this section.

3.6.1   Language Modeling

Like the image classification tasks in computer vision, language modeling is one of the most fundamental tasks in NLP. The most frequently studied and applied task on NLP associated with NAS is language modeling. Given a sequence of tokens, language modeling aims at predicting the probability of its appearance. Most of the state-of-the-art models on NLP tasks have the same pattern: A backbone network like ELMo [132], BERT [90], GPT [78], etc. that is pretrained on large language modeling tasks, and a classification network that is fine-tuned on the target tasks with the pretrained backbone. Even the fine-tune process can be removed from the training pipeline in recent NLP frameworks [78]. The pretrained backbone network pushes the frontier of almost all kinds of NLP tasks to a new level [90], which demonstrate the importance of language modeling to NLP area. Recently, a large amount of researches focus on applying NAS to language modeling tasks. The searched network can also be transferred as a backbone to solve other tasks such as named entity recognition [107, 113].

The first NAS for language modeling framework is proposed by Zoph et al. [158]. They design a fixed tree structured recurrent architecture search space, with the leaf nodes being the last hidden states h_{t-1} and current inputs x_t. Then, each node in the tree needs two operations to merge and transform its inputs. Each edge in the tree represents a linear transformation. The output of the previous state c_{t-1} is also considered and certain cell injection operations are searched for injection of c_{t-1} and output of c_t. An RNN-based controller is designed to sample the recurrent architectures for training, whose performance is used as reward to update the controller's parameters. The overall NAS framework is shown in Fig. 3.35.

Pham et al. [133] describe the search space of the recurrent architecture using DAG (see Fig. 3.36). Each node selects one input from the previous nodes, transforms it by a linear transformation and then applies a selected activation operation to derive the output. This kind of search space design allows topology search in addition to operation search in [158] and also enable weight sharing among different topology, which largely reduces time and memory costs. To improve the performance of the model while training stably, skip connection is also applied on every transformation of nodes. Like [158], they use reinforcement learning to search for the best architecture. This kind of DAG space design dominates the research trends and is widely used [91, 107, 121, 123, 154]. Liu et al. [121] relax the search

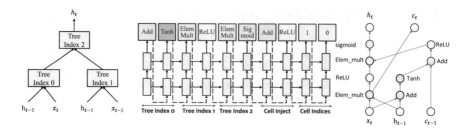

Fig. 3.35 An example of a recurrent cell constructed from a tree that has two leaf nodes (base 2) and one internal node in [158]. Left: the tree that defines the computation steps to be predicted by controller. Center: an example set of predictions made by the controller for each computation step in the tree. Right: the computation graph of the recurrent cell constructed from example predictions of the controller

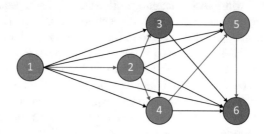

Fig. 3.36 DAG design of recurrent cells in [133]. The graph represents the entire search space while the red arrows define a model in the search space, which is decided by a controller. Here, node 1 is the input to the model whereas nodes 3 and 6 are the model's outputs

space with softmax and propose a differentiable search algorithm to solve this task. Jiang et al. [107] further remove the restriction that every two nodes have only one edge at most, and use softmax over all incoming edges to create a more flexible search space.

Li et al. [113] extend the search space of the recurrent cell. They formalize the definition of RNN into two parts: inter-cell network and intra-cell network (see Fig. 3.37). The inter-cell network aims at deriving the representation vector of previous cells and inputs:

$$\hat{h}_{t-1} = f(h_{[0,t-1]}; x_{[1,t-1]}),$$ (3.21)

$$\hat{x}_t = g(x_{[1,t]}; h_{[0,t-1]}).$$ (3.22)

The intra-cell network is designed to process the derived representation vectors of inter-cell network and output the hidden state of current step:

$$h_t = \pi(\hat{h}_{t-1}, \hat{x}_t),$$ (3.23)

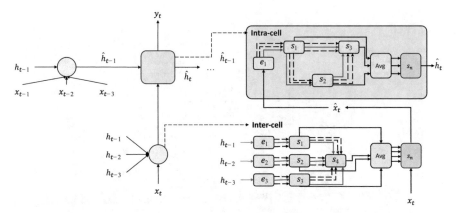

Fig. 3.37 An example of inter-cell and intra-cell network in [113]. h_t, x_t stands for hidden state and input at timestamp t. $\hat{h}_t$, $\hat{x}_t$ stands for the intermediate feature maps

where h, $\hat{h}$ represent the hidden states, x, $\hat{x}$ represent the inputs, f, g, π represent networks. For intra-cell network search space design, only a fixed number of historical hidden states and inputs are leveraged as input in order to keep the search space applicable and tractable. The network is further split into two sub-networks with their inputs separated according to the input types. Then the final hidden state of the current cell is output by Hadamard product of the outputs of the two subnets. For inter-cell network search space, they adopt the one used in [121, 133]. They use differentiable search framework to jointly search for the best architectures.

So et al. [138] propose the first space for Transformer [141] architectures. The Transformer-encoder consists of several stacked cells, each cell having several blocks (Fig. 3.38). Each block has two branch inputs, and each input is processed by independently selected norm function, dimension changing method and activation function. Then, the feature maps of the two branches are merged using a selected combiner function as the output of current blocks. All operations need to be selected and the number of units in the transformer encoder is searchable. The search space contains the original design of the transformer encoder. They use evolutionary algorithm to search, and allocate more training resources to get better performance through the dynamic hurdles.

3.6.2 Text Classification

Text classification is one of the basic task in NLP. The core problem is how to design a network structure that effectively captures the syntax and semantics hidden in texts and obtain a precise representation. Different from CV dominated by CNN, the SOTA structure of text representation is much more complex. In addition to the typical RNN, CNN is also widely used in this task. In addition, new structures

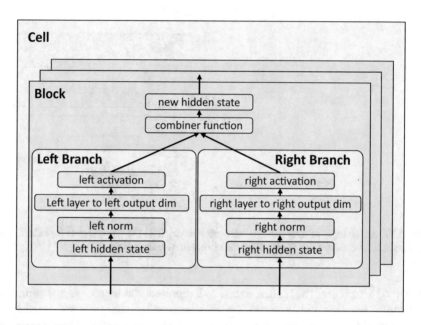

Fig. 3.38 Architecture composition from encoding in [138]. Each block produces a new hidden state that is added to the pool of hidden states subsequent blocks can select as branch inputs. Each encoder has 6 unique blocks per cell and each decoder has 8 unique blocks per cell. Each cell is repeated *number of cells* times

such as transformers have been proposed in recent years. The large number of available structures makes it difficult for researchers to design structures suitable for their tasks. However, as NAS has achieved success in many fields, it also provides opportunities for text classification tasks. In this field, research works usually focus on the design of search space. A good search space is the key to efficient search, and the design of search space is mainly based on people's prior knowledge.

Wang et al. [146] design a novel search space for text representation, which is called TextNAS. They find that CNN, RNN and Transformer[141] have their own advantages while layer mixture can lead to better performance than using only one layer. It is argued that the macro search space is better than micro search space in text representation scenarios. They improve the macro search space by letting it embody multi-path ensemble because different kindw of layers extract distinct features and an ensemble of them can provide better representationw. The macro search space in this work is described by a general DAG. A network instance is sampled by traversing the layers according to the topology order for the layers. In the search space, there are four commonly used kinds of layers as candidates, namely convolutional layer, recurrent layer, pooling layer and multi-head self-attention layer. These layers can stack freely without the change of shape of input tensor. Algorithm from [133] is adopted to search for the best models.

3.6.3   Other NLP Tasks

In addition to the natural language processing tasks mentioned above, there are many other natural language processing tasks that can be further improved by using NAS. So et al. [138] propose the first NAS application on neural machine translation tasks. They adopt a similar space design for Transformer-decoder, and add the decoder-specific *attend-to-encoder* operation to the primitive operation pool. The search algorithm is also the same as the one used in language modeling.

3.7   Other Applications of NAS

So et al. [75] introduce a method to automate the process of discovering optimization methods. Following [158], a RNN controller is used to generate an update equation for the optimizer. The generated update equation is based on a set of primitive functions such as the gradient, running average of the gradient. The training framework is based on reinforcement learning. The controller generates update rules, whose performance will be estimated by the child network and the accuracy on the validation set acts as the reward signal. Then, the performance is used to update the controller with policy gradients. The result shows that the searched update rules are better than many optimizers such as Adam, RMSProp or SGD.

Pérez-Rúa et al. [131] introduce NAS to solve the problem of finding architectures for multi-model classification models. A novel and generic search space is designed to extend a larger amount of fusion architectures. In detail, the method considers two modalities, i.e., there are two kinds of modalities in one sample. Formally, one sample can be denoted as $(x, y; z)$ where x is the first modality, y is the second one and z is the supervision label. Data fusion is introduced through a third neural network which can be seen in Fig. 3.39. The fusion layer combines three inputs: the output of the previous fusion layers and one output from each modality. This design produces large search space of fusion architectures. However, they don't exhaustively explore all the possibilities due to the huge computation complexity. They use Sequential Model-Based Optimization (SMBO) which progressively explores a search space by dividing it into complexity levels and provides well enough architectures.

Pasunuru et al. [129] adopt NAS for continual learning and multi-task learning. They study two kinds of task settings. One setting is multiple tasks arrive sequentially, which requires the model to adapt to new task while maintaining the performance of the previous learned tasks. They solve it by making the model parameters block sparse and constraining the new learned parameters to be orthogonal to the previous knowledge, that is, the weight parameters are not all shared, and the newly learned edges are always independent (see Fig. 3.40). The other setting is that multiple tasks arrive at the same time and the goal is to learn

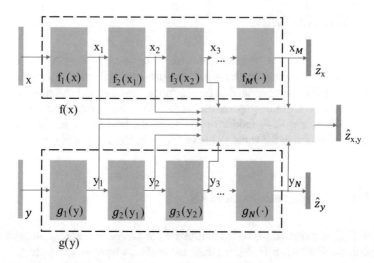

Fig. 3.39 General structure of a bi-modal fusion network in [131]. Top: a neural network with several hidden layers (grey boxes) with input **x**, and output $\hat{z}_x$. Bottom: a second network with input **y**, and output $\hat{z}_y$

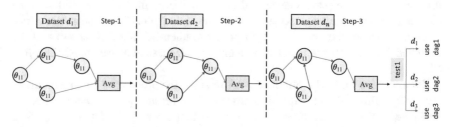

Fig. 3.40 Continual architecture search (CAS): blue edges (weight parameters) are shared, newly learned edges are represented with red edges

a single universal cell to adapt to all tasks and a new unseen task. They focus on finding a general cell structure which jointly performs well on all the tasks. The controller takes the performance of the sampled cell structure as a reward to update the new sample distribution (see Fig. 3.41). The search space is then designed accordingly. As for the search algorithm, they adapt ENAS [133] for both tasks.

Alletto et al. [71] aim to get the minimal human intervention required to deploy a AutoML method into real world scenarios especially when the modality changes. When a NAS method is evaluated, people often use the performance of the model that it finds as the only metric. This work proposes that the amount of human intervention during the deployment of the AutoML algorithm should be also considered because methods with similar accuracy may have great differences in the difficulty of applying to different modes. They propose a set of criteria to solve the difficulty of applying NAS methods into real world scenarios. This is the first fully automated NAS method for multi-model learning. In detail, the multimodel search

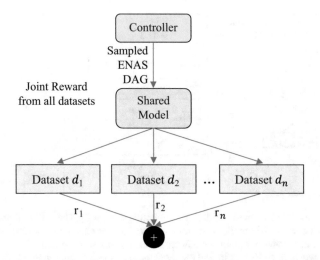

Fig. 3.41 Multi-task cell structure learning using joint rewards from n datasets

space is divided into two parts—feature extractors and fusion network. They adopt the micro-architecture search paradigm, which is to obtain the final architecture by stacking cells. The graph structure and operations set from ENAS [133] are adopted to design the cells. They improve the fusion network from [131] by allowing a fusion layer connect to any cell at a lower or equal depth.

Kang et al. [108] adopt NAS to address the inherent model capacity issue between teacher network and student network towards knowledge distillation. The goal of knowledge distillation is to transfer knowledge from one model to another. They think the limited capacity in the student network is the bottleneck of knowledge distillation. As a novel method, they abandon the assumption that student network structure is fixed. Instead, they incorporate NAS with knowledge distillation so as to obtain sufficient capacity. In addition, operations are searched to be added after each stage so as to enable the searched student network have higher precision potential than the teacher network.

Negrinho et al. [127] propose a formal language for encoding search space over general computational graphs. The language is designed to decouple the implementation of the search space and the search algorithm. The operation to be searched can be represented in a simple and consistent way. It is a useful tool for researchers to design their search space. The language is designed in a modular manner, which makes it have better programmability and reusability. Users can use the language constructed to encode search space in the literature. The encoded search space can be easily used by search algorithms in a consistent interface, which makes it convenient for researchers to compare different search spaces and search algorithms.

Wang et al. [145] propose a joint search method for network architecture, pruning and quantization polic named APQ. The three parts are usually separately optimized

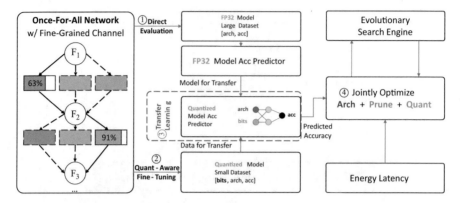

Fig. 3.42 An overview of the joint design methodology in [145]. The serial number represents the order of the steps. The model first trains an accuracy predictor for the full precision NN, then incrementally trains an accuracy predictor for the quantized NN (predictor-transfer). Finally, evolutionary search is performed to find the specialized NN architecture with quantization policy that fits hardware constraints

in the previous methods and this work searches the three parts in a joint manner. The overall framework is shown in Fig. 3.42. There is a once-for-all network with fine-grained channels, an accuracy predictor and evolution search to jointly optimize the three parts. In Fig. 3.42, the sequence numbers represent the order of the steps. The accuracy predictor for the precision neural network is firstly trained, then the predictor is incrementally trained for the quantized neural network. Finally, the evolutionary search is performed to find the specialized neural network architecture which fits hardware constraints. The joint search space is much larger than the original setting. Different from other works, the sampled network is not trained to obtain the actual accuracy because of the high time consumption. Instead, they deal with this by using the quantization-aware accuracy predictor. Firstly, they generate a large dataset of neural network architecture labeled by ImageNet accuracy. Then these data are used to train an accuracy predictor without quantization. Finally, the weights are transferred to a quantization-aware predictor. This method greatly reduces the time consumption.

Geifman et al. [98] propose to use NAS for active learning tasks. They depict the network search space using $(B, N_{blocks}, N_{stacks})$ and N_{stacks}, where B is block type that is fixed, N_{blocks} is the number of blocks in one stack and N_{stacks} is the number of stacks in the whole network. N_{blocks} and N_{stacks} are searchable. In order to deal with the active learning situation, they use iNAS to progressively grow the networks, where in each grow step the input at (B, i, j), the $(B, i+1, j)$ and $(B, [\frac{ij}{j+1}]+1, j+1)$ are trained and evaluated to compare with the orignial input. The model with best performance will be selected as output. The grow step is applied repeatedly until the output is the same as input or maximum setting of network is reached. They perform grow step repeatedly every time new data is received to progressively increase the network capacity for large datasets.

3.7.1   NAS for Network Compression

Network compression refers to given a well-trained model, how to compress it for better inference efficiency or smaller computation/memory costs. Before the birth of NAS, the network compression was completed by some hand-crafted rules, which requires experts in corresponding field to try multiple times to design the rule. Because of the huge exploration space, hand-crafted rules are always suboptimal. Nowadays, many researchers attempt to use NAS to alleviate the difficulty of network compression design, and they achieve better results.

Ashok et al. [74] is the first paper that utilizes NAS on general network compression task. They propose a two-stage network compression framework and use reinforcement learning to select the best compression method. At the first layer removal stage, every layer of the teacher network is considered whether to remove or not. At the second layer shrinkage stage, the compression ratio for every configurable variable is determined in each layer. The above decisions are made by a bilinear RNN agent that has the same input length as the total number of layers of the teacher network. A reward function is designed to reward models with high compression ratio while maintaining high performance.

Following [74], Cao et al. [81] also apply NAS on network compression task. Different from [74] which uses reinforcement learning to derive the best architecture, Cao et al. utilize Bayesian optimization to do the optimization. They describe the derivation of reward function in [74] as a Gaussian process, and use kernel function to depict the probability of the performance of given architectures. To derive the kernel function, they propose to use Bi-LSTM to learn the architecture embedding space, which is similar to [74] but can support skip connections. To solve the problem of maximization of acquisition function, they propose a random search method to randomly manipulate current models. In addition to the layer removal and shrinkage in [74], they also include adding skip connections. They utilize multi-kernel strategy to prevent overfitting problem. Bayesian optimization is used to find the best compressed model.

An et al. [72] also utilize NAS to do compression for their Photonet. They only consider whether to keep certain component of their network or not, and derive a 31-bit search space. They use knowledge distillation, together with the perceptual loss of child network and network compression rate as their search objectives. Evolutionary algorithm is used as the search strategy of compressed task.

Fu et al. [96] propose the first NAS network compression framework on GANs. Different from [137], they utilize NAS to compress the GAN instead of just relying on network pruning. Because of knowledge extraction based on teacher network, the framework does not need the Gan discriminator which is usually discarded after full training. To allow task specific GAN compression, they separate the tasks solved by GANs into two groups: image translation and super resolution. They build a macro architecture for each task (see Fig. 3.43), and leave some cells to be searched by NAS. Each searchable cell represents an operation that can be selected from a primitive operation pool. They allow searching the width of each cell by selecting an

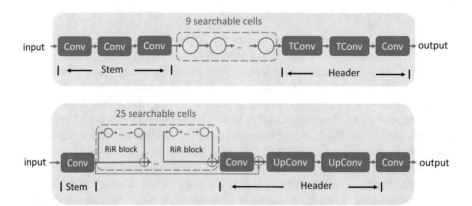

Fig. 3.43 Macro architectures for image translation (top) and super resolution (bottom) designed in [96]

expansion ratio. For the search algorithm, softmax is used to relax the search space selected by the operator, and Gumbel softmax is used to sample the expansion ratio to save memory and computational overhead. Gradient descent is used to directly optimize the loss function composed of distillation loss and compression ratio.

Chen et al. [83] apply NAS to search binary neural architectures. Although the given teacher model is not compressed in this paper, we also list it here because Binary Neural Network (BNN) is an important method to reduce the memory and computing cost. Similar to [121], they depict the search space of image classification network using DAGs. The difference is that all CNN operations in [121] are also binarized durning training. In order to speed up the search speed, they adopt memory efficient PC-DARTS algorithm [150], combined with the performance-based operation space reduction, which reduces the search space and improves the search efficiency.

3.7.2   NAS for Quick Adaption on New Tasks

Usually, NAS needs a lot of time and computing resources to find an optimal architecture for a new task. It is suboptimal to directly borrow the searched architecture of on old tasks. How to quickly apply NAS on new tasks becomes a new problem. In order to solve this problem, researchers have made some attempts by using multi task learning, transfer learning and meta learning.

Wong et al. [147] propose to combine transfer learning with multi-task training to guide the search for new tasks. The basic NAS framework is borrowed from [158], with a slight change of controller—a task-specific embedding vector is concatenated to every input of the controllers to sample networks condition on given task. The reward of different tasks are normalized to have the same average and variance to normalize training of different tasks. They use multi-task training to train the controllers. For a given new task, a task-specific embedding is randomly initialized and is updated together with the controller weights during searching process. The pretrained parameters in multi-task training can be regarded as guidance of successive training on new tasks.

Shaw et al. [136] combine meta learning with NAS for the first time. Following [159], they also design normal cells and reduction cells to construct the whole network. Each cell contains several layers, while each layer decides which previous layers are the inputs and what operations to use. To enable meta-learning, they form the NAS problem in a Bayesian manner and derive the corresponding loss function using Evidence Lower BOund (ELBO) of Maximum Log-likelihood Estimation (MLE). They also use optimization embedding to describe the posterior distribution of model and architecture parameters condition on datasets and minimize the objective on several tasks. Therefore, for a new task, model and architecture parameters can be quickly sampled.

Lian et al. [115] propose a Transferable Neural Architecture Search framework (T-NAS) that leverages meta learning to enable quick adaptation on new tasks. Different from the Bayesian inference framework in [136], they utilize Model-Agnostic Meta-Learning (MAML) [95] to allow quick adaptation. What is different from MAML is that they also assign parameters to architectures in addition to the parameters of weights. The meta-architectures are also optimized during meta-training process with the training of meta-weights. To speed up training, they adopt simultaneous gradient descent for both the meta-weights and meta-architectures.

Similar to [115], Wang et al. [143] propose a NAS framework named M-NAS based on gradient-based meta-learning. Different from [115], where the architectures are parameterized by independent meta-architectures, they use a recurrent autoencoder to derive the task-specific representation. The task-specific representation is then processed by a Multi-Layer Perceptron (MLP) to generate task specific architectures. Meta weights and task specific modulator are designed to generate task specific initial weights. Then, the task specific model and architectures can be quickly optimized within a few gradient descents. The parameters of autoencoder, MLP, meta weight and modulator are jointly optimized in an end-to-end fashion.

Morgado et al. [125] apply NAS to quickly adjust the architectures and weights of pre-trained teacher networks on the given tasks. The NetTailor method firstly augments the pre-trained network with some task-specific low-complexity proxy layers as shown in Fig. 3.44. After that, the constructed supernetwork is directly trained on given tasks. The layers with low impact on model performance will be pruned and the remaining network is fine-tuned to derive the final model. For the augmentation of teacher network, each layer will be augmented with a set of lean proxy layers that introduce a skip connection between the current layer and all previous layers (see Fig. 3.45). Every new layer, together with the original pretrained layer, will be assigned with an impact score and the new outputs of current layer is the weighted sum of all augmented layers and original layer. The new super-network

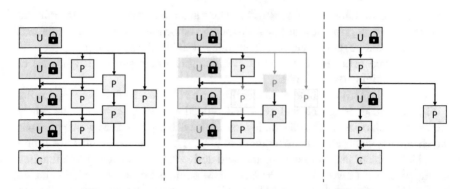

Fig. 3.44 Architecture fine-tuning with NETTAILOR [125]. Pre-trained blocks shown in gray, task-specific in yellow. Left: pre-trained CNN augmented with low-complexity blocks that introduce skip connections. Center: optimization prunes blocks of poor trade-off between complexity and impact on recognition. Right: the final network is a combination of pre-trained and task-specific blocks

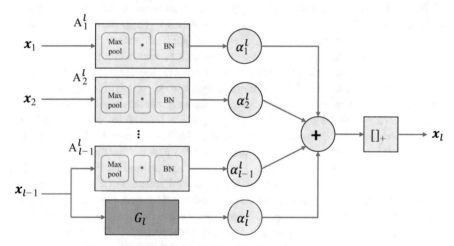

Fig. 3.45 Augmentation of pre-trained block at one layer G_l with multiple proxy layers A_p^l in [125], $\mathbf{x}_i$ represents the network activation after layer i

is then trained using gradient descent to minimize the loss function considering knowledge distillation loss and model complexity.

3.7.3   NAS for Out-of-Distribution Detection

Ardywibowo et al. [73] apply NAS on Out-of-Distribution (OoD) detection tasks. Unlike common application practices of NAS, they don't search for a single best architecture, but maintain the distribution of architectures. They maximize the

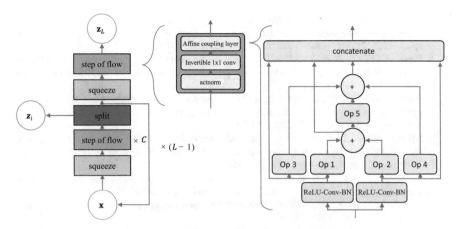

Fig. 3.46 Search space in [73] of a single block design following [109]

Widely Applicable Information Criteria (WAIC) score to improve model uncertainty quantification and OoD detection. The architecture based on invertible flow has attracted much attention due to its high efficiency and versatility. They build a layer-wise search space with a predefined macro-architecture, which consists of several stacked blocks that can have different architectures. Following [109], the micro search space design of block is shown in Fig. 3.46. The connection topology is fixed and only the operations need to be searched. Similar to [91], they relax the discrete search space, use Gumbel-Softmax to sample networks and update the shared weight and architecture distribution probability.

References

1. Wenwu Zhu, Xin Wang and Wen Gao. Multimedia intelligence: When multimedia meets artificial intelligence. *IEEE Transactions on Multimedia*, 22(7):1823–1835, 2020.
2. G. A. Carpenter, S. Grossberg, and J. H. Reynolds. Artmap: supervised real-time learning and classification of nonstationary data by a self-organizing neural network. In *[1991 Proceedings] IEEE Conference on Neural Networks for Ocean Engineering*, pages 341–342, 1991.
3. A. Nickabadi, M. M. Ebadzadeh, and R. Safabakhsh. Evaluating the performance of dnpso in dynamic environments. In *IEEE International Conference on Systems*, 2008.
4. Konstantin Shvachko, Hairong Kuang, Sanjay Radia, and Robert Chansler. The hadoop distributed file system. In *2010 IEEE 26th symposium on mass storage systems and technologies (MSST)*, pages 1–10. Ieee, 2010.
5. Simon Chan, Philip Treleaven, and Licia Capra. Continuous hyperparameter optimization for large-scale recommender systems. In *2013 IEEE International Conference on Big Data*, pages 350–358. IEEE, 2013.
6. Yehuda Koren. Factorization meets the neighborhood: a multifaceted collaborative filtering model. In *Proceedings of the 14th ACM SIGKDD international conference on Knowledge discovery and data mining*, pages 426–434, 2008.

7. Yunhong Zhou, Dennis Wilkinson, Robert Schreiber, and Rong Pan. Large-scale parallel collaborative filtering for the netflix prize. In *International conference on algorithmic applications in management*, pages 337–348. Springer, 2008.
8. Pawel Matuszyk, Renê Tatua Castillo, Daniel Kottke, and Myra Spiliopoulou. A comparative study on hyperparameter optimization for recommender systems. In *Workshop on Recommender Systems and Big Data Analytics*, pages 13–21, 2016.
9. Ananto Setyo Wicaksono and Ahmad Afif Supianto. Hyper parameter optimization using genetic algorithm on machine learning methods for online news popularity prediction. *Int. J. Adv. Comput. Sci. Appl*, 9(12):263–267, 2018.
10. Hugo Caselles-Dupré, Florian Lesaint, and Jimena Royo-Letelier. Word2vec applied to recommendation: Hyperparameters matter. In *Proceedings of the 12th ACM Conference on Recommender Systems*, pages 352–356, 2018.
11. Ian Dewancker, Michael McCourt, and Scott Clark. Bayesian optimization for machine learning: A practical guidebook. *arXiv preprint arXiv:1612.04858*, 2016.
12. Yves F Atchadé. A computational framework for empirical bayes inference. *Statistics and computing*, 21(4):463–473, 2011.
13. Madeleine Udell, Corinne Horn, Reza Zadeh, and Stephen Boyd. Generalized low rank models. *arXiv preprint arXiv:1410.0342*, 2014.
14. Arabin Kumar Dey, Raghav Somani, and Sreangsu Acharyya. A case study of empirical bayes in a user-movie recommendation system. *Communications in Statistics: Case Studies, Data Analysis and Applications*, 3(1–2):1–6, 2017.
15. Bruno Giovanni Galuzzi, Ilaria Giordani, Antonio Candelieri, Riccardo Perego, and Francesco Archetti. Bayesian optimization for recommender system. In *World Congress on Global Optimization*, pages 751–760. Springer, 2019.
16. Shubham Gautam, Pallavi Gupta, Satyajit Swain, and Rohit Ranjan. Discovery oriented model selection in music recommendation.
17. Guangxiang Zeng, Hengshu Zhu, Qi Liu, Ping Luo, Enhong Chen, and Tong Zhang. Matrix factorization with scale-invariant parameters. In *Twenty-Fourth International Joint Conference on Artificial Intelligence*, 2015.
18. Francis Bach, Julien Mairal, and Jean Ponce. Convex sparse matrix factorizations. *arXiv preprint arXiv:0812.1869*, 2008.
19. Emmanuel J Candes and Yaniv Plan. Matrix completion with noise. *Proceedings of the IEEE*, 98(6):925–936, 2010.
20. Elad Hazan. Sparse approximate solutions to semidefinite programs. In *Latin American symposium on theoretical informatics*, pages 306–316. Springer, 2008.
21. Martin Jaggi and Marek Sulovskỳ. A simple algorithm for nuclear norm regularized problems. In *ICML*, 2010.
22. Lidan Wang, Minwei Feng, Bowen Zhou, Bing Xiang, and Sridhar Mahadevan. Efficient hyper-parameter optimization for nlp applications. In *Proceedings of the 2015 Conference on Empirical Methods in Natural Language Processing*, pages 2112–2117, 2015.
23. Yelp dataset challenge. http://www.yelp.com/dataset_challenge/.
24. Yasser Ganjisaffar, Rich Caruana, and Cristina Videira Lopes. Bagging gradient-boosted trees for high precision, low variance ranking models. In *Proceedings of the 34th international ACM SIGIR conference on Research and development in Information Retrieval*, pages 85–94, 2011.
25. Jasper Snoek, Hugo Larochelle, and Ryan P Adams. Practical bayesian optimization of machine learning algorithms. *arXiv preprint arXiv:1206.2944*, 2012.
26. Dani Yogatama and Noah A Smith. Bayesian optimization of text representations. *arXiv preprint arXiv:1503.00693*, 2015.
27. Katharina Eggensperger, Matthias Feurer, Frank Hutter, James Bergstra, Jasper Snoek, Holger Hoos, and Kevin Leyton-Brown. Towards an empirical foundation for assessing bayesian optimization of hyperparameters. In *NIPS workshop on Bayesian Optimization in Theory and Practice*, volume 10, page 3, 2013.

28. Rong-En Fan, Kai-Wei Chang, Cho-Jui Hsieh, Xiang-Rui Wang, and Chih-Jen Lin. Liblinear: A library for large linear classification. *the Journal of machine Learning research*, 9:1871–1874, 2008.

29. Chih-Jen Lin, Ruby C Weng, and S Sathiya Keerthi. Trust region newton method for large-scale logistic regression. *Journal of Machine Learning Research*, 9(4), 2008.

30. Eric S Tellez, Daniela Moctezuma, Sabino Miranda-Jiménez, and Mario Graff. An automated text categorization framework based on hyperparameter optimization. *Knowledge-Based Systems*, 149:110–123, 2018.

31. Pulkit Mehndiratta and Devpriya Soni. Identification of sarcasm using word embeddings and hyperparameters tuning. *Journal of Discrete Mathematical Sciences and Cryptography*, 22(4):465–489, 2019.

32. Ziyu Wang, Babak Shakibi, Lin Jin, and Nando Freitas. Bayesian multi-scale optimistic optimization. In *Artificial Intelligence and Statistics*, pages 1005–1014. PMLR, 2014.

33. Aditya Parameswaran, Hector Garcia-Molina, and Anand Rajaraman. Towards the web of concepts: Extracting concepts from large datasets. *Proceedings of the VLDB Endowment*, 3(1-2):566–577, 2010.

34. J-D Kim, Tomoko Ohta, Yuka Tateisi, and Juníchi Tsujii. Genia corpus—a semantically annotated corpus for bio-textmining. *Bioinformatics*, 19(suppl_1):i180–i182, 2003.

35. Li Deng, Geoffrey Hinton, and Brian Kingsbury. New types of deep neural network learning for speech recognition and related applications: An overview. In *2013 IEEE international conference on acoustics, speech and signal processing*, pages 8599–8603. IEEE, 2013.

36. George E Dahl, Tara N Sainath, and Geoffrey E Hinton. Improving deep neural networks for lvcsr using rectified linear units and dropout. In *2013 IEEE international conference on acoustics, speech and signal processing*, pages 8609–8613. IEEE, 2013.

37. Yoshua Bengio, Nicolas Boulanger-Lewandowski, and Razvan Pascanu. Advances in optimizing recurrent networks. In *2013 IEEE International Conference on Acoustics, Speech and Signal Processing*, pages 8624–8628. IEEE, 2013.

38. James Bergstra and Yoshua Bengio. Random search for hyper-parameter optimization. *Journal of machine learning research*, 13(2), 2012.

39. Colin Raffel and Daniel PW Ellis. Optimizing dtw-based audio-to-midi alignment and matching. In *2016 IEEE International Conference on Acoustics, Speech and Signal Processing (ICASSP)*, pages 81–85. IEEE, 2016.

40. Colin Raffel and Daniel PW Ellis. Large-scale content-based matching of midi and audio files. In *ISMIR*, pages 234–240, 2015.

41. Thierry Bertin-Mahieux, Daniel PW Ellis, Brian Whitman, and Paul Lamere. The million song dataset. 2011.

42. Erik Bochinski, Tobias Senst, and Thomas Sikora. Hyper-parameter optimization for convolutional neural network committees based on evolutionary algorithms. In *2017 IEEE International Conference on Image Processing (ICIP)*, pages 3924–3928. IEEE, 2017.

43. Dounia Lakhmiri, Sébastien Le Digabel, and Christophe Tribes. Hypernomad: Hyperparameter optimization of deep neural networks using mesh adaptive direct search. *arXiv preprint arXiv:1907.01698*, 2019.

44. Sébastien Le Digabel. *NOMAD: Nonlinear optimization with the MADS algorithm*. Groupe d'études et de recherche en analyse des décisions, 2010.

45. Li Deng. The mnist database of handwritten digit images for machine learning research [best of the web]. *IEEE Signal Processing Magazine*, 29(6):141–142, 2012.

46. Alex Krizhevsky, Geoffrey Hinton, et al. Learning multiple layers of features from tiny images. 2009.

47. Jean-François Connolly, Eric Granger, and Robert Sabourin. An adaptive classification system for video-based face recognition. *Information Sciences*, 192:50–70, 2012.

48. Samira Ebrahimi Kahou, Xavier Bouthillier, Pascal Lamblin, Caglar Gulcehre, Vincent Michalski, Kishore Konda, Sébastien Jean, Pierre Froumenty, Yann Dauphin, Nicolas Boulanger-Lewandowski, et al. Emonets: Multimodal deep learning approaches for emotion recognition in video. *Journal on Multimodal User Interfaces*, 10(2):99–111, 2016.

49. Muhammad Usman Yaseen, Ashiq Anjum, Omer Rana, and Nikolaos Antonopoulos. Deep learning hyper-parameter optimization for video analytics in clouds. *IEEE Transactions on Systems, Man, and Cybernetics: Systems*, 49(1):253–264, 2018.
50. Jakub Król and Michał Grega. Optimization of configuration parameter set in video analysis. *Journal of Image and Graphics*, 7(4), 2019.
51. Khaing Suu Htet and Myint Myint Sein. Market intelligence analysis on age estimation and gender classification on events with deep learning hyperparameters optimization and sdn controllers. In *2020 IEEE Conference on Computer Applications (ICCA)*, pages 1–5. IEEE, 2020.
52. Onur Dikmen and A Taylan Cemgil. Gamma markov random fields for audio source modeling. *IEEE Transactions on Audio, Speech, and Language Processing*, 18(3):589–601, 2009.
53. Rune Johan Borgli, Håkon Kvale Stensland, Michael Alexander Riegler, and Pål Halvorsen. Automatic hyperparameter optimization for transfer learning on medical image datasets using bayesian optimization. In *2019 13th International Symposium on Medical Information and Communication Technology (ISMICT)*, pages 1–6. IEEE, 2019.
54. Stefano Alletto, Shenyang Huang, Vincent Francois-Lavet, Yohei Nakata, and Guillaume Rabusseau. RandomNet: Towards Fully Automatic Neural Architecture Design for Multimodal Learning. 2020.
55. Jingyuan Chen, Hanwang Zhang, Xiangnan He, Liqiang Nie, Wei Liu, and Tat-Seng Chua. Attentive collaborative filtering: Multimedia recommendation with item-and component-level attention. In *Proceedings of the 40th International ACM SIGIR conference on Research and Development in Information Retrieval*, pages 335–344, 2017.
56. Yashar Deldjoo, Markus Schedl, Paolo Cremonesi, Gabriella Pasi, and PAOLO CRE-MONESI. Content-based multimedia recommendation systems: definition and application domains. 2018.
57. Jacob Devlin, Ming-Wei Chang, Kenton Lee, and Kristina Toutanova. Bert: Pre-training of deep bidirectional transformers for language understanding. *arXiv preprint arXiv:1810.04805*, 2018.
58. Kaiming He, Xiangyu Zhang, Shaoqing Ren, and Jian Sun. Deep residual learning for image recognition. In *Proceedings of the IEEE conference on computer vision and pattern recognition*, pages 770–778, 2016.
59. Alexios Kotsifakos, Panagiotis Papapetrou, Jaakko Hollmén, Dimitrios Gunopulos, and Vassilis Athitsos. A survey of query-by-humming similarity methods. In *Proceedings of the 5th International Conference on PErvasive Technologies Related to Assistive Environments*, pages 1–4, 2012.
60. G. Li, Z. Ming, H. Li, and T.S. Chua. Early versus Late Fusion in Semantic Video Analysis. *Proceedings of the seventeen ACM international conference on Multimedia*, pages 773–776, 2009.
61. Xirong Li, Chaoxi Xu, Gang Yang, Zhineng Chen, and Jianfeng Dong. W2vv++ fully deep learning for ad-hoc video search. In *Proceedings of the 27th ACM International Conference on Multimedia*, pages 1786–1794, 2019.
62. Hanxiao Liu, Karen Simonyan, and Yiming Yang. DARTS: Differentiable architecture search. *arXiv preprint arXiv:1806.09055*, 2018.
63. Mohammad Norouzi and David J Fleet. Minimal loss hashing for compact binary codes. In *ICML*, 2011.
64. Juan Manuel Perez-Rua, Valentin Vielzeuf, Stephane Pateux, Moez Baccouche, and Frederic Jurie. MFAS: Multimodal fusion architecture search. *Proceedings of the IEEE Computer Society Conference on Computer Vision and Pattern Recognition*, 2019-June:6959–6968, 2019.
65. Hieu Pham, Melody Y Guan, Barret Zoph, Quoc V Le, and Jeff Dean. Efficient neural architecture search via parameter sharing. *arXiv preprint arXiv:1802.03268*, 2018.
66. Xin Wang, Wenwu Zhu, and Chenghao Liu. Semi-supervised deep quantization for cross-modal search. In *Proceedings of the 27th ACM International Conference on Multimedia*, pages 1730–1739, 2019.

67. Yinwei Wei, Xiang Wang, Liqiang Nie, Xiangnan He, Richang Hong, and Tat-Seng Chua. Mmgcn: Multi-modal graph convolution network for personalized recommendation of micro-video. In *Proceedings of the 27th ACM International Conference on Multimedia*, pages 1437–1445, 2019.

68. Xiaodong Yang, Pavlo Molchanov, and Jan Kautz. Multilayer and multimodal fusion of deep neural networks for video classification. *MM 2016 - Proceedings of the 2016 ACM Multimedia Conference*, pages 978–987, 2016.

69. Zhou Yu, Yuhao Cui, Jun Yu, Meng Wang, Dacheng Tao, and Qi Tian. Deep Multimodal Neural Architecture Search. 2020.

70. Barret Zoph and Quoc V Le. Neural architecture search with reinforcement learning. *arXiv preprint arXiv:1611.01578*, 2016.

71. Stefano Alletto, Shenyang Huang, Vincent François-Lavet, Yohei Nakata, and Guillaume Rabusseau. Randomnet: Towards fully automatic neural architecture design for multimodal learning. *CoRR*, abs/2003.01181, 2020.

72. Jie An, Haoyi Xiong, Jun Huan, and Jiebo Luo. Ultrafast photorealistic style transfer via neural architecture search. In *AAAI*, pages 10443–10450, 2020.

73. Randy Ardywibowo, Shahin Boluki, Xinyu Gong, Zhangyang Wang, and Xiaoning Qian. NADS: neural architecture distribution search for uncertainty awareness. *CoRR*, abs/2006.06646, 2020.

74. Anubhav Ashok, Nicholas Rhinehart, Fares Beainy, and Kris M. Kitani. N2N learning: Network to network compression via policy gradient reinforcement learning. In *6th International Conference on Learning Representations, ICLR 2018, Vancouver, BC, Canada, April 30 - May 3, 2018, Conference Track Proceedings*. OpenReview.net, 2018.

75. Irwan Bello, Barret Zoph, Vijay Vasudevan, and Quoc V. Le. Neural optimizer search with reinforcement learning. In Doina Precup and Yee Whye Teh, editors, *Proceedings of the 34th International Conference on Machine Learning, ICML 2017, Sydney, NSW, Australia, 6–11 August 2017*, volume 70 of *Proceedings of Machine Learning Research*, pages 459–468. PMLR, 2017.

76. Gabriel Bender, Hanxiao Liu, Bo Chen, Grace Chu, Shuyang Cheng, Pieter-Jan Kindermans, and Quoc V Le. Can weight sharing outperform random architecture search? an investigation with tunas. In *Proceedings of the IEEE/CVF Conference on Computer Vision and Pattern Recognition*, pages 14323–14332, 2020.

77. Andrew Brock, Theodore Lim, James M. Ritchie, and Nick Weston. SMASH: one-shot model architecture search through hypernetworks. In *6th International Conference on Learning Representations, ICLR 2018, Vancouver, BC, Canada, April 30 - May 3, 2018, Conference Track Proceedings*. OpenReview.net, 2018.

78. Tom B. Brown, Benjamin Mann, Nick Ryder, Melanie Subbiah, Jared Kaplan, Prafulla Dhariwal, Arvind Neelakantan, Pranav Shyam, Girish Sastry, Amanda Askell, Sandhini Agarwal, Ariel Herbert-Voss, Gretchen Krueger, Tom Henighan, Rewon Child, Aditya Ramesh, Daniel M. Ziegler, Jeffrey Wu, Clemens Winter, Christopher Hesse, Mark Chen, Eric Sigler, Mateusz Litwin, Scott Gray, Benjamin Chess, Jack Clark, Christopher Berner, Sam McCandlish, Alec Radford, Ilya Sutskever, and Dario Amodei. Language models are few-shot learners. 2020.

79. Han Cai, Jiacheng Yang, Weinan Zhang, Song Han, and Yong Yu. Path-level network transformation for efficient architecture search. In Jennifer G. Dy and Andreas Krause, editors, *Proceedings of the 35th International Conference on Machine Learning, ICML 2018, Stockholmsmässan, Stockholm, Sweden, July 10–15, 2018*, volume 80 of *Proceedings of Machine Learning Research*, pages 677–686. PMLR, 2018.

80. Han Cai, Ligeng Zhu, and Song Han. Proxylessnas: Direct neural architecture search on target task and hardware. In *7th International Conference on Learning Representations, ICLR 2019, New Orleans, LA, USA, May 6–9, 2019*. OpenReview.net, 2019.

81. Shengcao Cao, Xiaofang Wang, and Kris M. Kitani. Learnable embedding space for efficient neural architecture compression. In *7th International Conference on Learning Representations, ICLR 2019, New Orleans, LA, USA, May 6–9, 2019*. OpenReview.net, 2019.

82. Bo Chen, Golnaz Ghiasi, Hanxiao Liu, Tsung-Yi Lin, Dmitry Kalenichenko, Hartwig Adam, and Quoc V Le. Mnasfpn: Learning latency-aware pyramid architecture for object detection on mobile devices. In *Proceedings of the IEEE/CVF Conference on Computer Vision and Pattern Recognition*, pages 13607–13616, 2020.

83. Hanlin Chen, Li'an Zhuo, Baochang Zhang, Xiawu Zheng, Jianzhuang Liu, David S. Doermann, and Rongrong Ji. Binarized neural architecture search. In *The Thirty-Fourth AAAI Conference on Artificial Intelligence, AAAI 2020, The Thirty-Second Innovative Applications of Artificial Intelligence Conference, IAAI 2020, The Tenth AAAI Symposium on Educational Advances in Artificial Intelligence, EAAI 2020, New York, NY, USA, February 7–12, 2020*, pages 10526–10533. AAAI Press, 2020.

84. Liang-Chieh Chen, Maxwell D. Collins, Yukun Zhu, George Papandreou, Barret Zoph, Florian Schroff, Hartwig Adam, and Jonathon Shlens. Searching for efficient multi-scale architectures for dense image prediction. In Samy Bengio, Hanna M. Wallach, Hugo Larochelle, Kristen Grauman, Nicolò Cesa-Bianchi, and Roman Garnett, editors, *Advances in Neural Information Processing Systems 31: Annual Conference on Neural Information Processing Systems 2018, NeurIPS 2018, 3–8 December 2018, Montréal, Canada*, pages 8713–8724, 2018.

85. Liang-Chieh Chen, George Papandreou, Iasonas Kokkinos, Kevin Murphy, and Alan L Yuille. Deeplab: Semantic image segmentation with deep convolutional nets, atrous convolution, and fully connected crfs. *IEEE transactions on pattern analysis and machine intelligence*, 40(4):834–848, 2017.

86. Liang-Chieh Chen, George Papandreou, Florian Schroff, and Hartwig Adam. Rethinking atrous convolution for semantic image segmentation. *CoRR*, abs/1706.05587, 2017.

87. Sheng Chen, Yang Liu, Xiang Gao, and Zhen Han. Mobilefacenets: Efficient cnns for accurate real-time face verification on mobile devices. In Jie Zhou, Yunhong Wang, Zhenan Sun, Zhenhong Jia, Jianjiang Feng, Shiguang Shan, Kurban Ubul, and Zhenhua Guo, editors, *Biometric Recognition - 13th Chinese Conference, CCBR 2018, Urumqi, China, August 11–12, 2018, Proceedings*, volume 10996 of *Lecture Notes in Computer Science*, pages 428–438. Springer, 2018.

88. Yukang Chen, Gaofeng Meng, Qian Zhang, Shiming Xiang, Chang Huang, Lisen Mu, and Xinggang Wang. Renas: Reinforced evolutionary neural architecture search. In *Proceedings of the IEEE Conference on computer vision and pattern recognition*, pages 4787–4796, 2019.

89. Jiequan Cui, Pengguang Chen, Ruiyu Li, Shu Liu, Xiaoyong Shen, and Jiaya Jia. Fast and practical neural architecture search. In *Proceedings of the IEEE International Conference on Computer Vision*, pages 6509–6518, 2019.

90. Jacob Devlin, Ming-Wei Chang, Kenton Lee, and Kristina Toutanova. BERT: pre-training of deep bidirectional transformers for language understanding. In Jill Burstein, Christy Doran, and Thamar Solorio, editors, *Proceedings of the 2019 Conference of the North American Chapter of the Association for Computational Linguistics: Human Language Technologies, NAACL-HLT 2019, Minneapolis, MN, USA, June 2–7, 2019, Volume 1 (Long and Short Papers)*, pages 4171–4186. Association for Computational Linguistics, 2019.

91. Xuanyi Dong and Yi Yang. Searching for a robust neural architecture in four gpu hours. In *Proceedings of the IEEE Conference on computer vision and pattern recognition*, pages 1761–1770, 2019.

92. Jiemin Fang, Yuzhu Sun, Kangjian Peng, Qian Zhang, Yuan Li, Wenyu Liu, and Xinggang Wang. Fast neural network adaptation via parameter remapping and architecture search. In *8th International Conference on Learning Representations, ICLR 2020, Addis Ababa, Ethiopia, April 26–30, 2020*. OpenReview.net, 2020.

93. Jiemin Fang, Yuzhu Sun, Qian Zhang, Yuan Li, Wenyu Liu, and Xinggang Wang. Densely connected search space for more flexible neural architecture search. In *Proceedings of the IEEE/CVF Conference on Computer Vision and Pattern Recognition*, pages 10628–10637, 2020.

94. Clément Farabet, Camille Couprie, Laurent Najman, and Yann LeCun. Learning hierarchical features for scene labeling. *IEEE Trans. Pattern Anal. Mach. Intell.*, 35(8):1915–1929, 2013.

95. Chelsea Finn, Pieter Abbeel, and Sergey Levine. Model-agnostic meta-learning for fast adaptation of deep networks. In Doina Precup and Yee Whye Teh, editors, *Proceedings of the 34th International Conference on Machine Learning, ICML 2017, Sydney, NSW, Australia, 6–11 August 2017*, volume 70 of *Proceedings of Machine Learning Research*, pages 1126–1135. PMLR, 2017.

96. Yonggan Fu, Wuyang Chen, Haotao Wang, Haoran Li, Yingyan Lin, and Zhangyang Wang. Autogan-distiller: Searching to compress generative adversarial networks. *CoRR*, abs/2006.08198, 2020.

97. Chen Gao, Yunpeng Chen, Si Liu, Zhenxiong Tan, and Shuicheng Yan. Adversarialnas: Adversarial neural architecture search for gans. In *Proceedings of the IEEE/CVF Conference on Computer Vision and Pattern Recognition*, pages 5680–5689, 2020.

98. Yonatan Geifman and Ran El-Yaniv. Deep active learning with a neural architecture search. In Hanna M. Wallach, Hugo Larochelle, Alina Beygelzimer, Florence d'Alché-Buc, Emily B. Fox, and Roman Garnett, editors, *Advances in Neural Information Processing Systems 32: Annual Conference on Neural Information Processing Systems 2019, NeurIPS 2019, 8–14 December 2019, Vancouver, BC, Canada*, pages 5974–5984, 2019.

99. Golnaz Ghiasi, Tsung-Yi Lin, and Quoc V Le. Nas-fpn: Learning scalable feature pyramid architecture for object detection. In *Proceedings of the IEEE conference on computer vision and pattern recognition*, pages 7036–7045, 2019.

100. Daniel Golovin, Benjamin Solnik, Subhodeep Moitra, Greg Kochanski, John Karro, and D. Sculley. Google vizier: A service for black-box optimization. In *Proceedings of the 23rd ACM SIGKDD International Conference on Knowledge Discovery and Data Mining, Halifax, NS, Canada, August 13–17, 2017*, pages 1487–1495. ACM, 2017.

101. Xinyu Gong, Shiyu Chang, Yifan Jiang, and Zhangyang Wang. Autogan: Neural architecture search for generative adversarial networks. In *Proceedings of the IEEE International Conference on Computer Vision*, pages 3224–3234, 2019.

102. Dazhou Guo, Dakai Jin, Zhuotun Zhu, Tsung-Ying Ho, Adam P Harrison, Chun-Hung Chao, Jing Xiao, and Le Lu. Organ at risk segmentation for head and neck cancer using stratified learning and neural architecture search. In *Proceedings of the IEEE/CVF Conference on Computer Vision and Pattern Recognition*, pages 4223–4232, 2020.

103. Jianyuan Guo, Kai Han, Yunhe Wang, Chao Zhang, Zhaohui Yang, Han Wu, Xinghao Chen, and Chang Xu. Hit-detector: Hierarchical trinity architecture search for object detection. In *Proceedings of the IEEE/CVF Conference on Computer Vision and Pattern Recognition*, pages 11405–11414, 2020.

104. Zichao Guo, Xiangyu Zhang, Haoyuan Mu, Wen Heng, Zechun Liu, Yichen Wei, and Jian Sun. Single path one-shot neural architecture search with uniform sampling. *CoRR*, abs/1904.00420, 2019.

105. Roxana Istrate, Florian Scheidegger, Giovanni Mariani, Dimitrios Nikolopoulos, Constantine Bekas, and Adelmo Cristiano Innocenza Malossi. Tapas: Train-less accuracy predictor for architecture search. In *Proceedings of the AAAI Conference on Artificial Intelligence*, volume 33, pages 3927–3934, 2019.

106. Chenhan Jiang, Shaoju Wang, Xiaodan Liang, Hang Xu, and Nong Xiao. Elixirnet: Relation-aware network architecture adaptation for medical lesion detection. In *AAAI*, pages 11093–11100, 2020.

107. Yufan Jiang, Chi Hu, Tong Xiao, Chunliang Zhang, and Jingbo Zhu. Improved differentiable architecture search for language modeling and named entity recognition. In *Proceedings of the 2019 Conference on Empirical Methods in Natural Language Processing and the 9th International Joint Conference on Natural Language Processing (EMNLP-IJCNLP)*, pages 3576–3581, 2019.

108. Minsoo Kang, Jonghwan Mun, and Bohyung Han. Towards oracle knowledge distillation with neural architecture search. In *AAAI*, pages 4404–4411, 2020.

109. Durk P Kingma and Prafulla Dhariwal. Glow: Generative flow with invertible 1x1 convolutions. In S. Bengio, H. Wallach, H. Larochelle, K. Grauman, N. Cesa-Bianchi, and R. Garnett, editors, *Advances in Neural Information Processing Systems 31*, pages 10215–10224. Curran Associates, Inc., 2018.

110. Guohao Li, Guocheng Qian, Itzel C Delgadillo, Matthias Muller, Ali Thabet, and Bernard Ghanem. Sgas: Sequential greedy architecture search. In *Proceedings of the IEEE/CVF Conference on Computer Vision and Pattern Recognition*, pages 1620–1630, 2020.

111. Wei Li, Shaogang Gong, and Xiatian Zhu. Neural graph embedding for neural architecture search. In *AAAI*, pages 4707–4714, 2020.

112. Xin Li, Yiming Zhou, Zheng Pan, and Jiashi Feng. Partial order pruning: for best speed/accuracy trade-off in neural architecture search. In *Proceedings of the IEEE Conference on computer vision and pattern recognition*, pages 9145–9153, 2019.

113. Yinqiao Li, Chi Hu, Yuhao Zhang, Nuo Xu, Yufan Jiang, Tong Xiao, Jingbo Zhu, Tongran Liu, and Changliang Li. Learning architectures from an extended search space for language modeling. In Dan Jurafsky, Joyce Chai, Natalie Schluter, and Joel R. Tetreault, editors, *Proceedings of the 58th Annual Meeting of the Association for Computational Linguistics, ACL 2020, Online, July 5–10, 2020*, pages 6629–6639. Association for Computational Linguistics, 2020.

114. Zhihang Li, Teng Xi, Jiankang Deng, Gang Zhang, Shengzhao Wen, and Ran He. Gpnas: Gaussian process based neural architecture search. In *Proceedings of the IEEE/CVF Conference on Computer Vision and Pattern Recognition*, pages 11933–11942, 2020.

115. Dongze Lian, Yin Zheng, Yintao Xu, Yanxiong Lu, Leyu Lin, Peilin Zhao, Junzhou Huang, and Shenghua Gao. Towards fast adaptation of neural architectures with meta learning. In *International Conference on Learning Representations*, 2019.

116. Peiwen Lin, Peng Sun, Guangliang Cheng, Sirui Xie, Xi Li, and Jianping Shi. Graph-guided architecture search for real-time semantic segmentation. In *Proceedings of the IEEE/CVF Conference on Computer Vision and Pattern Recognition*, pages 4203–4212, 2020.

117. Tsung-Yi Lin, Priya Goyal, Ross B. Girshick, Kaiming He, and Piotr Dollár. Focal loss for dense object detection. In *IEEE International Conference on Computer Vision, ICCV 2017, Venice, Italy, October 22–29, 2017*, pages 2999–3007. IEEE Computer Society, 2017.

118. Chenxi Liu, Liang-Chieh Chen, Florian Schroff, Hartwig Adam, Wei Hua, Alan L Yuille, and Li Fei-Fei. Auto-deeplab: Hierarchical neural architecture search for semantic image segmentation. In *Proceedings of the IEEE conference on computer vision and pattern recognition*, pages 82–92, 2019.

119. Chenxi Liu, Barret Zoph, Maxim Neumann, Jonathon Shlens, Wei Hua, Li-Jia Li, Li Fei-Fei, Alan Yuille, Jonathan Huang, and Kevin Murphy. Progressive neural architecture search. In *Proceedings of the European Conference on Computer Vision (ECCV)*, pages 19–34, 2018.

120. Hanxiao Liu, Karen Simonyan, Oriol Vinyals, Chrisantha Fernando, and Koray Kavukcuoglu. Hierarchical representations for efficient architecture search. In *6th International Conference on Learning Representations, ICLR 2018, Vancouver, BC, Canada, April 30 - May 3, 2018, Conference Track Proceedings*. OpenReview.net, 2018.

121. Hanxiao Liu, Karen Simonyan, and Yiming Yang. DARTS: differentiable architecture search. In *7th International Conference on Learning Representations, ICLR 2019, New Orleans, LA, USA, May 6–9, 2019*. OpenReview.net, 2019.

122. Wei Liu, Dragomir Anguelov, Dumitru Erhan, Christian Szegedy, Scott E. Reed, Cheng-Yang Fu, and Alexander C. Berg. SSD: single shot multibox detector. In Bastian Leibe, Jiri Matas, Nicu Sebe, and Max Welling, editors, *Computer Vision - ECCV 2016 - 14th European Conference, Amsterdam, The Netherlands, October 11–14, 2016, Proceedings, Part I*, volume 9905 of *Lecture Notes in Computer Science*, pages 21–37. Springer, 2016.

123. Renqian Luo, Fei Tian, Tao Qin, Enhong Chen, and Tie-Yan Liu. Neural architecture optimization. In Samy Bengio, Hanna M. Wallach, Hugo Larochelle, Kristen Grauman, Nicolò Cesa-Bianchi, and Roman Garnett, editors, *Advances in Neural Information Processing Systems 31: Annual Conference on Neural Information Processing Systems 2018, NeurIPS 2018, 3–8 December 2018, Montréal, Canada*, pages 7827–7838, 2018.

124. Jieru Mei, Yingwei Li, Xiaochen Lian, Xiaojie Jin, Linjie Yang, Alan L. Yuille, and Jianchao Yang. Atomnas: Fine-grained end-to-end neural architecture search. In *8th International Conference on Learning Representations, ICLR 2020, Addis Ababa, Ethiopia, April 26–30, 2020*. OpenReview.net, 2020.

125. Pedro Morgado and Nuno Vasconcelos. Nettailor: Tuning the architecture, not just the weights. In *Proceedings of the IEEE Conference on Computer Vision and Pattern Recognition*, pages 3044–3054, 2019.

126. Niv Nayman, Asaf Noy, Tal Ridnik, Itamar Friedman, Rong Jin, and Lihi Zelnik-Manor. XNAS: neural architecture search with expert advice. In Hanna M. Wallach, Hugo Larochelle, Alina Beygelzimer, Florence d'Alché-Buc, Emily B. Fox, and Roman Garnett, editors, *Advances in Neural Information Processing Systems 32: Annual Conference on Neural Information Processing Systems 2019, NeurIPS 2019, 8–14 December 2019, Vancouver, BC, Canada*, pages 1975–1985, 2019.

127. Renato Negrinho, Matthew Gormley, Geoffrey J Gordon, Darshan Patil, Nghia Le, and Daniel Ferreira. Towards modular and programmable architecture search. In *Advances in Neural Information Processing Systems*, pages 13715–13725, 2019.

128. Vladimir Nekrasov, Hao Chen, Chunhua Shen, and Ian Reid. Fast neural architecture search of compact semantic segmentation models via auxiliary cells. In *Proceedings of the IEEE Conference on computer vision and pattern recognition*, pages 9126–9135, 2019.

129. Ramakanth Pasunuru and Mohit Bansal. Continual and multi-task architecture search. In Anna Korhonen, David R. Traum, and Lluís Màrquez, editors, *Proceedings of the 57th Conference of the Association for Computational Linguistics, ACL 2019, Florence, Italy, July 28- August 2, 2019, Volume 1: Long Papers*, pages 1911–1922. Association for Computational Linguistics, 2019.

130. Wei Peng, Xiaopeng Hong, Haoyu Chen, and Guoying Zhao. Learning graph convolutional network for skeleton-based human action recognition by neural searching. In *AAAI*, pages 2669–2676, 2020.

131. Juan-Manuel Pérez-Rúa, Valentin Vielzeuf, Stéphane Pateux, Moez Baccouche, and Frédéric Jurie. Mfas: Multimodal fusion architecture search. In *Proceedings of the IEEE Conference on computer vision and pattern recognition*, pages 6966–6975, 2019.

132. Matthew E. Peters, Mark Neumann, Mohit Iyyer, Matt Gardner, Christopher Clark, Kenton Lee, and Luke Zettlemoyer. Deep contextualized word representations. In Marilyn A. Walker, Heng Ji, and Amanda Stent, editors, *Proceedings of the 2018 Conference of the North American Chapter of the Association for Computational Linguistics: Human Language Technologies, NAACL-HLT 2018, New Orleans, Louisiana, USA, June 1–6, 2018, Volume 1 (Long Papers)*, pages 2227–2237. Association for Computational Linguistics, 2018.

133. Hieu Pham, Melody Y. Guan, Barret Zoph, Quoc V. Le, and Jeff Dean. Efficient neural architecture search via parameter sharing. In Jennifer G. Dy and Andreas Krause, editors, *Proceedings of the 35th International Conference on Machine Learning, ICML 2018, Stockholmsmässan, Stockholm, Sweden, July 10–15, 2018*, volume 80 of *Proceedings of Machine Learning Research*, pages 4092–4101. PMLR, 2018.

134. Michael S. Ryoo, A. J. Piergiovanni, Mingxing Tan, and Anelia Angelova. Assemblenet: Searching for multi-stream neural connectivity in video architectures. In *8th International Conference on Learning Representations, ICLR 2020, Addis Ababa, Ethiopia, April 26–30, 2020*. OpenReview.net, 2020.

135. Mark Sandler, Andrew G. Howard, Menglong Zhu, Andrey Zhmoginov, and Liang-Chieh Chen. Mobilenetv2: Inverted residuals and linear bottlenecks. In *2018 IEEE Conference on Computer Vision and Pattern Recognition, CVPR 2018, Salt Lake City, UT, USA, June 18–22, 2018*, pages 4510–4520. IEEE Computer Society, 2018.

136. Albert Shaw, Wei Wei, Weiyang Liu, Le Song, and Bo Dai. Meta architecture search. In Hanna M. Wallach, Hugo Larochelle, Alina Beygelzimer, Florence d'Alché-Buc, Emily B. Fox, and Roman Garnett, editors, *Advances in Neural Information Processing Systems 32: Annual Conference on Neural Information Processing Systems 2019, NeurIPS 2019, 8–14 December 2019, Vancouver, BC, Canada*, pages 11225–11235, 2019.

137. Han Shu, Yunhe Wang, Xu Jia, Kai Han, Hanting Chen, Chunjing Xu, Qi Tian, and Chang Xu. Co-evolutionary compression for unpaired image translation. In *2019 IEEE/CVF International Conference on Computer Vision, ICCV 2019, Seoul, Korea (South), October 27 - November 2, 2019*, pages 3234–3243. IEEE, 2019.

138. David R. So, Quoc V. Le, and Chen Liang. The evolved transformer. In Kamalika Chaudhuri and Ruslan Salakhutdinov, editors, *Proceedings of the 36th International Conference on Machine Learning, ICML 2019, 9–15 June 2019, Long Beach, California, USA*, volume 97 of *Proceedings of Machine Learning Research*, pages 5877–5886. PMLR, 2019.

139. Masanori Suganuma, Mete Ozay, and Takayuki Okatani. Exploiting the potential of standard convolutional autoencoders for image restoration by evolutionary search. In Jennifer G. Dy and Andreas Krause, editors, *Proceedings of the 35th International Conference on Machine Learning, ICML 2018, Stockholmsmässan, Stockholm, Sweden, July 10–15, 2018*, volume 80 of *Proceedings of Machine Learning Research*, pages 4778–4787. PMLR, 2018.

140. Mingxing Tan, Bo Chen, Ruoming Pang, Vijay Vasudevan, Mark Sandler, Andrew Howard, and Quoc V Le. Mnasnet: Platform-aware neural architecture search for mobile. In *Proceedings of the IEEE Conference on Computer Vision and Pattern Recognition*, pages 2820–2828, 2019.

141. Ashish Vaswani, Noam Shazeer, Niki Parmar, Jakob Uszkoreit, Llion Jones, Aidan N Gomez, Łukasz Kaiser, and Illia Polosukhin. Attention is all you need. In *Advances in neural information processing systems*, pages 5998–6008, 2017.

142. Alvin Wan, Xiaoliang Dai, Peizhao Zhang, Zijian He, Yuandong Tian, Saining Xie, Bichen Wu, Matthew Yu, Tao Xu, Kan Chen, et al. Fbnetv2: Differentiable neural architecture search for spatial and channel dimensions. In *Proceedings of the IEEE/CVF Conference on Computer Vision and Pattern Recognition*, pages 12965–12974, 2020.

143. Jiaxing Wang, Jiaxiang Wu, Haoli Bai, and Jian Cheng. M-nas: Meta neural architecture search. In *AAAI*, pages 6186–6193, 2020.

144. Ning Wang, Yang Gao, Hao Chen, Peng Wang, Zhi Tian, Chunhua Shen, and Yanning Zhang. Nas-fcos: Fast neural architecture search for object detection. In *Proceedings of the IEEE/CVF Conference on Computer Vision and Pattern Recognition*, pages 11943–11951, 2020.

145. Tianzhe Wang, Kuan Wang, Han Cai, Ji Lin, Zhijian Liu, Hanrui Wang, Yujun Lin, and Song Han. Apq: Joint search for network architecture, pruning and quantization policy. In *Proceedings of the IEEE/CVF Conference on Computer Vision and Pattern Recognition*, pages 2078–2087, 2020.

146. Yujing Wang, Yaming Yang, Yiren Chen, Jing Bai, Ce Zhang, Guinan Su, Xiaoyu Kou, Yunhai Tong, Mao Yang, and Lidong Zhou. Textnas: A neural architecture search space tailored for text representation. In *AAAI*, pages 9242–9249, 2020.

147. Catherine Wong, Neil Houlsby, Yifeng Lu, and Andrea Gesmundo. Transfer learning with neural automl. In Samy Bengio, Hanna M. Wallach, Hugo Larochelle, Kristen Grauman, Nicolò Cesa-Bianchi, and Roman Garnett, editors, *Advances in Neural Information Processing Systems 31: Annual Conference on Neural Information Processing Systems 2018, NeurIPS 2018, 3–8 December 2018, Montréal, Canada*, pages 8366–8375, 2018.

148. Bichen Wu, Xiaoliang Dai, Peizhao Zhang, Yanghan Wang, Fei Sun, Yiming Wu, Yuandong Tian, Peter Vajda, Yangqing Jia, and Kurt Keutzer. Fbnet: Hardware-aware efficient convnet design via differentiable neural architecture search. In *Proceedings of the IEEE Conference on Computer Vision and Pattern Recognition*, pages 10734–10742, 2019.

149. Hang Xu, Lewei Yao, Wei Zhang, Xiaodan Liang, and Zhenguo Li. Auto-fpn: Automatic network architecture adaptation for object detection beyond classification. In *Proceedings of the IEEE International Conference on Computer Vision*, pages 6649–6658, 2019.

150. Yuhui Xu, Lingxi Xie, Xiaopeng Zhang, Xin Chen, Guo-Jun Qi, Qi Tian, and Hongkai Xiong. PC-DARTS: partial channel connections for memory-efficient architecture search. In *8th International Conference on Learning Representations, ICLR 2020, Addis Ababa, Ethiopia, April 26–30, 2020*. OpenReview.net, 2020.

151. Lewei Yao, Hang Xu, Wei Zhang, Xiaodan Liang, and Zhenguo Li. SM-NAS: structural-to-modular neural architecture search for object detection. In *The Thirty-Fourth AAAI Conference on Artificial Intelligence, AAAI 2020, The Thirty-Second Innovative Applications*

of Artificial Intelligence Conference, IAAI 2020, The Tenth AAAI Symposium on Educational Advances in Artificial Intelligence, EAAI 2020, New York, NY, USA, February 7–12, 2020, pages 12661–12668. AAAI Press, 2020.

152. Qihang Yu, Dong Yang, Holger Roth, Yutong Bai, Yixiao Zhang, Alan L Yuille, and Daguang Xu. C2fnas: Coarse-to-fine neural architecture search for 3d medical image segmentation. In *Proceedings of the IEEE/CVF Conference on Computer Vision and Pattern Recognition*, pages 4126–4135, 2020.

153. Haokui Zhang, Ying Li, Hao Chen, and Chunhua Shen. Memory-efficient hierarchical neural architecture search for image denoising. In *Proceedings of the IEEE/CVF Conference on Computer Vision and Pattern Recognition*, pages 3657–3666, 2020.

154. Tunhou Zhang, Hsin-Pai Cheng, Zhenwen Li, Feng Yan, Chengyu Huang, Hai Helen Li, and Yiran Chen. Autoshrink: A topology-aware nas for discovering efficient neural architecture. In *AAAI*, pages 6829–6836, 2020.

155. Yiheng Zhang, Zhaofan Qiu, Jingen Liu, Ting Yao, Dong Liu, and Tao Mei. Customizable architecture search for semantic segmentation. In *Proceedings of the IEEE Conference on Computer Vision and Pattern Recognition*, pages 11641–11650, 2019.

156. Hengshuang Zhao, Jianping Shi, Xiaojuan Qi, Xiaogang Wang, and Jiaya Jia. Pyramid scene parsing network. In *2017 IEEE Conference on Computer Vision and Pattern Recognition, CVPR 2017, Honolulu, HI, USA, July 21–26, 2017*, pages 6230–6239. IEEE Computer Society, 2017.

157. Zhao Zhong, Junjie Yan, Wei Wu, Jing Shao, and Cheng-Lin Liu. Practical block-wise neural network architecture generation. In *Proceedings of the IEEE conference on computer vision and pattern recognition*, pages 2423–2432, 2018.

158. Barret Zoph and Quoc V. Le. Neural architecture search with reinforcement learning. In *5th International Conference on Learning Representations, ICLR 2017, Toulon, France, April 24–26, 2017, Conference Track Proceedings*. OpenReview.net, 2017.

159. Barret Zoph, Vijay Vasudevan, Jonathon Shlens, and Quoc V. Le. Learning transferable architectures for scalable image recognition. In *2018 IEEE Conference on Computer Vision and Pattern Recognition, CVPR 2018, Salt Lake City, UT, USA, June 18–22, 2018*, pages 8697–8710. IEEE Computer Society, 2018.

Chapter 4
Meta-Learning for Multimedia

Meta-learning, or learning to learn, is a sub-domain of machine-learning that aims to learn prior knowledge (i.e., meta-knowledge) across different learning tasks to improve the learning algorithm itself. In recent years, meta-learning has seen a drastic rise of interest especially in deep learning [1], since it provides potential value for tackling the conventional challenges in deep learning, e.g., data and computational bottleneck, generalization ability, etc. Specifically, meta-learning algorithms are most often applied to alleviate the data scarcity problem in the few-shot learning paradigm, where the size of dataset of each task is extremely small compared with the needs of deep models, and the meta-learner derives meta-knowledge across several learning tasks (or episodes) and rapidly adapts this knowledge to new tasks with limited amount of training data.

In multimedia context, meta-learning has great potential to be applied to various practical scenes. The motivation for leveraging meta-learning in multimedia problems mainly includes but is not limited to data efficiency (i.e., few-shot setting), generalization ability, and algorithm optimization. In this chapter, we will divide the applications of meta-learning in multimedia context into two application categories: meta-learning for multimedia search and recommendation, and meta-learning for vision and language. For the first category, we will discuss how meta-learning can be used in cold-start recommendation, recommender algorithm selection, and incremental product search. For the second category, we will introduce the application of meta-learning in classification, detection, image captioning, tracking, and visual question answering problems. We can summarize from this chapter that meta-learning serves as solutions with great potential in many multimedia applications, while more exploration is still needed in several promising future directions concluded in the subsections.

4.1   Meta-Learning for Multimedia Search and Recommendation

In Sects. 3.1 and 3.4 we have provided overviews on HPO and NAS for multimedia search and recommendation with a comprehensive discussion on the task characteristics and general frameworks of multimedia search and multimedia recommendation. In this section, we mainly focus on the application of another branch of advanced machine learning, i.e., meta-learning, on search or recommendation tasks. Note that although the following ideas are not intentionally designed for multimedia data, they are of significant potential value for multimedia search and recommendation applications since most of the meta-learning approaches are beyond specific search or recommendation models (i.e. model-agnostic). Therefore, one could simply replace the models (e.g., feature extractors or prediction models) with multimedia-related models to instantiate the meta-learning ideas.

Meta-learning, or *learning to learn*, focuses on deriving general knowledge (i.e., a prior) across different learning tasks, aiming to rapidly adapt to a new learning task with the prior and a small amount of training data [14]. The "prior" here is learned by a meta-learner during *meta-training* and has strong generalization capacity such that it can be easily and quickly adapted to the new tasks during *meta-testing*.

Meta-learning has recently been widely adopted under few-shot learning settings (e.g., few-shot image classification and few-shot reinforcement learning), where the data of the meta-testing tasks are scarce. We note that few-shot learning has been well studied [48] and here we will not go into details due to space limitation. Through extending few-shot learning settings to recommender systems, recent literature introduces meta-learning algorithms to solve the few-shot recommendation tasks (i.e., *cold-start recommendation*), which will be one focus of this section.

Another application of meta-learning for recommendation is the problem of *recommender algorithm selection* [3, 15], which will also be introduced in this section. In recommender systems, no single algorithm (e.g., Matrix Factorization, Collaborative Filtering, etc.) could consistently outperform other algorithms on every dataset. Therefore, to dynamically select the best algorithm(s) for a specific dataset to achieve overall high performance, scholars apply meta-learning ideas to learn the "prior" knowledge about how to select recommender algorithms based on dataset characteristics.

Less attention has been drawn to the meta-learning application for search (or retrieval) problems. In this section we briefly introduce a work on the task of few-shot incremental product search [24]. Finally, we summarize the section and attempt to provide future directions for applying meta-learning to multimedia scenarios. Note that the cutting-edge field of "multimodal meta-learning", as introduced in Sect. 2.4.2, is not related with the multiple modalities of multimedia data, but refers to the scenarios where the meta-learning tasks are of multiple categories, which is out of the scope of the discussion of this section.

4.1.1 Meta-Learning for Cold-Start Recommendation

As aforementioned, meta-learning approaches has shown strong power in few-shot settings. Since cold-start recommendation is the instantiation of few-shot learning problem in recommender systems, a natural and recent trend is to exploit the power of meta-learning in cold-start recommendation scenarios.

Before we start this topic, let us firstly introduce the problem setting and traditional solutions of cold-start recommendation. Generally, the cold-start recommendation problem [16] refers to the settings where the training data is scarce (e.g., new users, items, or scenarios with sparse user-item interactions) such that traditional collaborative-filtering-based algorithms fail to provide high-quality recommendation. As summarized in [14], traditional solutions basically rely on data augmentation by leveraging the side information of users or items, the high-order structures of user-item-object (e.g., item=movie and object=director) graphs, review texts and source domain data (by transfer learning).

While the above traditional solutions alleviate the cold-start problem at the data level, on the other hand, meta-learning solutions aim to alleviate the cold-start problem at the algorithm level, by learning the "prior" across different data-sparse recommendation tasks to guide the recommender learning in new tasks. Specifically, when we formulate cold-start recommendation as a meta-learning problem, each task is to learn the recommender for one cold-start user, item or scenario. Moreover, the "prior" here has different meanings in different meta-learning approaches (i.e., metric-based, memory-based or optimization-based as discussed in Sect. 2).

An overview of the literature on meta-learning for cold-start recommendation is demonstrated in Table 4.1. As shown in the table, bilevel optimization methods are the most popular meta-learning approach for cold-start recommendation, probably due to the generalization capacity of gradient-based optimization like MAML [106]. From the perspective of cold-start subject, most of the works define each task as learning the personalized recommender system for a specific user. Thus, as a starting example, let us firstly present more details about the bilevel optimization framework for cold-start user. From this example, we will later extend the frameworks for other cold-start object (cold-start item and scenario) and also other meta-learning approaches (memory-based and metric-based).

A. Bilevel Optimization for Cold-Start User

Bilevel optimization is a popular methodology in meta-learning literature [13, 19, 106]. The basic idea is to learn from meta-training tasks the general hyper-parameters of machine learning algorithms, including initialization (of model parameters), learning rate, stop policy, etc. To gradually find the best hyper-parameters to generalize to new tasks, bilevel optimization methods adopt two levels of gradient-based optimization: *the inner loop optimization (i.e. local update)* and *the outer loop optimization (i.e. global update)*. Specifically, each task in meta-

Table 4.1 Literature overview of meta-learning for cold-start recommendation

Paper	Cold-start subject or task in meta-learning	Methodology	Meta-learned mapping
LWA, NLBA [22]	User	Metric-based	User history -> model parameters
SEATLE [12]	Item	Metric-based	User, item, geographical info -> sample embedding
MetaEmb [17]	Item	Bilevel optimization	Ad feature -> Ad embeddinge
MeLU [11]	User	Bilevel optimization	User, item -> preference
MetaCS-DNN [2]	User	Bilevel optimization	User, item -> preference
MetaHIN [14]	User	Bilevel optimization	User, item, context graph -> preference
MAMO [4]	User	Bilevel optimization and Memory-based	User, item -> preference
s^2Meta [5]	Scenario	Bilevel optimization	User, item -> preference

learning divides its dataset into two parts: *support set* for training, which is often few-shot, and *query set* for testing. In each epoch of bilevel-optimization training, the algorithm firstly sample a batch of tasks, and locally update (i.e., train for a few steps) the (recommender) model on the support set of each task, which we call "inner loop optimization". Then, the algorithm calculates the (recommendation) losses on the query sets and globally update the hyper-parameters based on the sum of the query losses, which we call "outer loop optimization". As a final result, the hyper-parameters can easily adapt to the few-shot support sets of different meta-testing tasks and achieve high (recommendation) performance on the unlabeled query samples.

In cold-start user setting, the general motivation is to improve the personalized recommendation performance for new users, who have few historical user-item interactions. Since two new users with similar user profile (or friend structure) might receive totally the same recommendations when applied with content-based (or collaborative filtering) methods, scholars tend to pay more attention on personal interests by exploiting the user historical interactions. In meta-learning approaches, *learning the preferences of each user* is defined as a single task. The dataset of each task consists of the user's historically-interacted samples. In meta-training tasks (old users), the samples are randomly divided into support set and query set without intersection. In meta-testing tasks (new users), all the history samples belong to the support set, and the labels of future samples for recommendation are to be predicted.

The general framework of bilevel optimization for cold-start user setting is illustrated in Fig. 4.1. As we can see, the parameters of the recommender model f is denoted as θ, and the hyper-parameters for model learning is denoted as ω, which is regarded as the parameters of a "meta-learner" g. We leverage the bilevel algorithm described above to meta-learn a general ω, which could easily adapt to

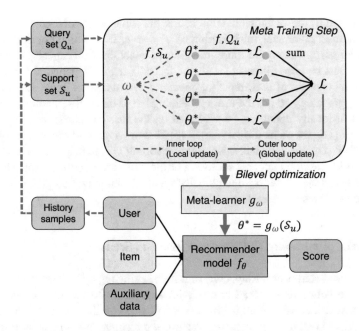

Fig. 4.1 A general framework of bilevel optimization for cold-start user setting

the support set $\mathcal{S}_u$ of a new user and infer the best recommender model parameters θ^* (i.e., $\theta^* = g_\omega(\mathcal{S}_u)$).

Let us now instantiate some representative examples of this framework. In MeLU [11], the recommender model f consists of an embedding module and a multi-layer perceptron (MLP), where the embedding module maps the user features (e.g., age and occupation) and item features (e.g., genre of movies) into embedding vectors, and the MLP maps the vector concatenation to the recommendation score output. The meta-learning algorithm leverages the bilevel optimization to learn an initialization of the parameters of f, which is an extension of MAML [106] to user preference estimation. Specifically, the inner loop locally updates the parameters of MLP, while the outer loop globally updates both the embedding module and the MLP. A parallel work MetaCS-DNN [2] slightly changes the inner loop to locally update both the embedding module and MLP.

For another example, MetaHIN [14] takes into account the Heterogeneous Information Networks (HIN) for data augmentation, combining the power of model level solution (meta-learning) and data level solution (HIN) to cold-start user problem. Intuitively, HIN refers to the heterogeneous graph of users, items (e.g., movies) and other objects (e.g., directors), and different semantic meta-paths (e.g., user-movie-director-movie) are defined as semantic-wise context for information aggregation. To this end, the recommender model $f = < g, h >$ firstly takes a context aggregator g to map all the history and context items of the user u to user embedding x_u, and then takes a MLP h to map user and item embedding concatenation to the recommendation score, the same as MeLU. Our aim now is to find the best parameter initialization of $f = < g, h >$ which could generalize to different users. To acquire more fine-grained generalization capacity to not only different users but also different "semantic facets" (i.e., meta-paths), the inner loop in MetaHIN consists of two steps: (a) *Semantic-wise adaption* firstly leverages the context aggregator g to get the user embedding based on each semantic meta-path, x_u^p. Next, it locally updates g to g^p based on the loss calculated with x_u^p on the support set (i.e., user's historical items). (b) *Task-wise adaption* then leverages the output of g^p (i.e., the semantic-wise user embedding $x_u^{p'}$) to help locally update the MLP h to h^p based on support set loss. For the outer loop optimization, MetaHIN finally fuses all the h^p and x_u^p and globally update the initialization of $f = < g, h >$ based on query loss.

B. Bilevel Optimization for Cold-Start Item and Scenario

Similar to cold-start user setting, cold-start item setting regards the recommendation for a single item i (e.g., an ad or a movie) as a task, and the support and query sets of each task consist of all the history interactions on this specific item. On the other hand, cold-start scenario setting regards the recommendation under a specific scenario c (e.g., baby store, travel store or new promotion) as a task, and the task dataset consists of all the user-item interactions that happens under this scenario. Intuitively, the basic difference between these two settings and cold-start user setting is the definition of a task in meta-learning, which decides different arrangement of samples. Therefore, it is natural that bilevel optimization could also be effective in these two settings.

A representative in cold-start item setting is MetaEmb [17], whose goal is to learn to map a new ad's features to an embedding, such that it could improve the accuracy of downstream click-through-rate (CTR) model. The meta-learning algorithm consists of two phases: (a) The *cold-start phase* calculates the CTR prediction loss (cross-entropy loss) l_a based on the initial embedding of an ad $\phi_i = h_w(x_i)$, where h_w is the mapping from ad feature x_i to embedding. (b) The *warm-up phase* then takes typical bilevel optimization by first locally updating ϕ_i on support set, calculating the loss l_b on query set and then globally updating the parameters w of h based on the fusion of l_a and l_b. In this way, the meta-learned initialization of mapping h_w should be both beneficial for both zero-shot brand-new ads (cold-start phase) and few-shot ads (warm-up phase).

Another work s^2Meta [5] takes the setting of cold-start scenario. The difference between it and previous bilevel optimization is that it considers not only initialization, but also the update policy and the stop policy. To be specific, an *update controller* modeled by LSTM [8] is adopted to predict the learning rate and momentum for the current local update step. Moreover, an *stop controller* modeled by REINFORCE [25] is also adopted to decide whether the current local update step should be stopped to avoid overfitting. Finally, the hyper-parameters, i.e., initialization, update controller and stop controller, are globally updated based

Table 4.2 Bilevel optimization meta-learning for cold start recommendation

Paper	Recommender model (f_θ)	Meta-learned hyperparas (ω)	Local update	Global update
MeLU [11]	Embedding + MLP	Initialization	MLP	All
MetaCS-DNN [2]	Embedding + MLP	Initialization	All	All
MetaHIN [14]	Context aggregation + MLP	Initialization	All (semantic- and task-wise)	All
MetaEmb [17]	Embedding + CTR model	Initialization	Embedding	Embedding
s^2Meta [5]	Embedding + MLP	Initialization, update policy, stop policy	MLP	MLP init., update and stop controller

on query loss. In this way, the training process under each new scenario could be flexibly adjusted according to the support set data.

We summarize the aforementioned 5 papers of bilevel optimization for cold-start recommendation in Table 4.2 for the sake of clearer presentation.

C. Memory-Based Meta-Learning for Cold-Start Recommendation

Despite the success of bilevel optimization meta-learning on cold-start recommendation, the existing methods suffer from instability, slow convergence and weak generalization. In particular, as argued in [4], the global initialization for all users (taking cold-start user setting as example) is not suitable to the users who show different gradient descent directions comparing with the majority of users in the training set. Therefore, a recent work Memory-Augmented Meta-Optimization (MAMO) [4] introduces memory-based meta-learning techniques to accelerate the bilevel optimization meta-training process and provide personalized initialization according to user profile.

Analogous to previous work, the recommender model f also consists of embedding modules (θ_u, θ_i) to respectively map user profile and item features to embeddings, and a MLP module (θ_r) to predict the recommendation score. The core design is that the memory mechanism is introduced to both of the modules. To begin with, *Feature-specific memory* is designed to map user profile p_u to a personalized bias term b_u, with the help of user embedding memory M_U and the profile memory M_P, both of which are 2D matrix. Specifically, we first calculate the attention values $a_u = attention(p_u, M_P)$. Then, b_u is obtained by $b_u = a_u^\top M_U$ and then used as an offset for user embedding: $\theta_u = \phi_u - \tau b_u$, where ϕ_u is the global parameter initialization for user embedding module and τ is a predefined scalar. In this way, we could provide a personalized bias term when initializing the model parameters and achieve better generalization capacity. In addition, *Task-specific memory* $M_{U,I}$

(a 3D cube) is also defined to provide user preference matrix $M_{u,I} = a_u^\top M_{U,I}$ (a 2D matrix), which serves as fast weights to transform the user and item embeddings in recommender model. To be specific, the final recommendation prediction is defined by $\text{MLP}_{\theta_r}(M_{u,I} \cdot [e_u, e_i])$, where $[e_u, e_i]$ denotes the concatenation of user and item embeddings. Finally, both the feature- and task-specific memories are globally updated in the fashion of Neural Turing Machine [6] based on query loss, while other updates are the same as bilevel optimization framework.

D. Metric-Based Meta-Learning for Cold-Start Recommendation

Let us conclude this subsection by introducing metric-based meta-learning [10, 21, 23], another popular methodology for few-shot learning, into cold-start recommendation. Intuitively, metric-based methods first takes an embedding module to map each support or query sample into the same embedding space. The goal is to meta-learn an embedding module across tasks such that for each task, the positive query samples are much closer (often measured by cosine similarity) to the embedding center of support set than negative query samples.

To introduce how to apply metric-based meta-learning into cold-start recommendation settings, we take SEATLE [12] as an example. The problem setting is "new user prediction": given a business (i.e., a point-of-interest or site on the map) in location-based social network (LBSN) applications like Yelp and Instagram, the system should rank all the new users to the business (who have not arrived at this site before) for recommendation. To this end, the authors adopt the cold-start item setting and randomly select K observed check-in tuples (business-user pair such that the user have already arrived at this specific business) as support set ("References"). Other c observed check-in tuples and c fake check-in tuples are sampled as query set.

To apply metric-based meta-learning, an embedding module, consisting of an initial embedding and self-attention layers, is defined to map all the support and query samples to the same embedding space. Then, we update the embedding module by ranking loss: $l = \max\{0, \text{sim}(R, q^-) - \text{sim}(R, q^+) + \gamma\}$, such that the similarity between reference embedding center R and positive embedding q^+ should be larger than that of negative embedding q^-. Finally, the recommendation prediction can be finished by recommending the closest query samples (new users) to the support samples of specific cold-start businesses. The framework of SEATLE is illustrated in Fig. 4.2.

To sum up, in Sect. 4.1.1, we discuss how to apply different meta-learning approaches, including bilevel optimization, memory-based and metric-based, to various cold-start recommendation settings (cold-start user, item or scenario).

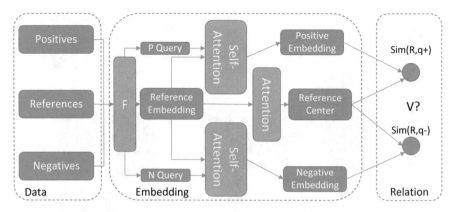

Fig. 4.2 The SEATLE framework [12], which applies metric-based meta-learning to cold-start item recommendation

4.1.2   Meta-Learning for Algorithm Selection

In the previous subsection, we have introduced how to meta-learn the *hyper-parameters of specific learning algorithm* (e.g., training a recommender model through SGD) for varying tasks. On the other hand, we could also meta-learn *how to select the most suitable algorithm(s)* (i.e., recommender model(s) with the corresponding learning strategies) for a specific recommendation task (dataset). The motivation here is not to alleviate cold-start problems any more. Instead, the intuition is that no single recommender algorithm could consistently outperform other algorithms on every dataset (or for every user) due to the heterogeneous data distributions [15]. To find the best algorithm(s), a naive idea is to apply each algorithm on each dataset and select the best, which is computational expensive. Therefore, scholars attempt to leverage meta-learning ideas to find the "prior" knowledge about algorithm selection across various recommendation tasks and adapt the prior to new tasks to achieve better recommendation performance.

Let us first define the problem of recommender algorithm (model) selection from the meta-learning perspective [3], illustrated in Fig. 4.3. To begin with, we have a meta-dataset, each meta-example of which is a recommendation dataset consisting of training (support) and testing (query) data. Our target is to meta-learn an algorithm selector, the input of which is the meta-features (or characteristics) of each dataset. Then, the algorithm selector outputs the absolute performance (e.g., error, accuracy, etc.) or relative performance (i.e., a ranking) of a set of algorithms on a specific dataset, based on which we could select the best one or several algorithms for this dataset.

Traditionally, literature focuses on how to define meta-features of datasets which "effectively describe how strongly a dataset matches the bias of each algorithm" and could help the algorithm selector to better predict the absolute performance of each algorithm. Generally, meta-features could be divided into three groups [3].

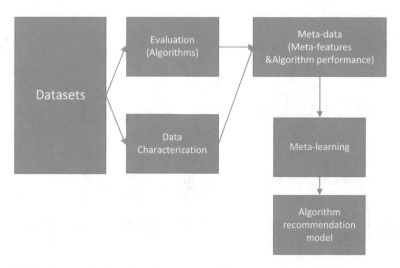

Fig. 4.3 The framework of traditional meta-learning for recommendation algorithm selection [3]

(a) *Statistical and information-theoretical* meta-features describe the dataset by its patterns of statistics (e.g., the number of examples, skewness, etc.) and information theory (e.g., entropy, mutual information, etc.). (b) *Model-based* meta-features are properties extracted from models induced from the dataset (e.g., the number of leaf nodes in decision trees). (c) *Landmarkers* are fast estimates (by simple models) of the algorithm performance on specific dataset. Based on these meta-features, traditional methods model the meta-learning goal as a performance estimation problem, adopting regression algorithms (e.g., linear regression) as meta-learner, or as a best algorithm prediction problem, adopting classification algorithms (e.g., SVM, kNN, etc.) as meta-learner. As for the recommender algorithm set for selection, the most popular algorithms includes heuristics, matrix factorization (MF) and neural networks (NN). We refer to [3] for more details.

Recently, the power of bilevel optimization based meta-learning (MAML [106]) has also been exploited for recommender model selection to achieve end-to-end training, circumventing manual definition of meta-features. A representative is MetaSelector [15], which focuses on user-level adaptive recommender model selection. Specifically, each dataset for model selection belongs to a specific user, which is the same as the setting of cold-start user recommendation. The goal is then to select, for each individual user, the best model (or model combination) from a given set of base models (e.g., FM [18], FFM [9], DeepFM [7], etc.). A model selector (e.g., MLP) is designed to take as input the data feature and the predictions of all base models, and then outputs a distribution on the base models, which is a kind of "soft model selection". The summation of base model predictions weighted by this soft selection is regarded as the final output of the current model combination and is taken as evidence for bilevel optimization. To be more specific, the meta-training process of MetaSelector algorithm is as follows. Step 1: pretrain all the base models on the whole meta-training datasets (users). Step 2: For each epoch we sample a batch of users, and do the bilevel optimization. That is, for each of the users u, we first locally update both the parameters of model selector ϕ and all based models $\theta = (\theta_1, \ldots, \theta_K)$ based on the loss on corresponding support set. Then we leverage the updated model selector ϕ^u and base models θ^u to calculate the loss on each query set and finally globally update the initialization of ϕ and θ. In such a way, we could automatically train a model selector to select the best model combination based on specific individual user (dataset) without any manual designed meta-features.

4.1.3 Meta-Learning for Incremental Product Search

The main difference between search and recommendation frameworks is the prediction module. In recommendation models, the prediction module often takes a flexible model (e.g., MLP) to map the user and item embeddings to a recommendation score. On the other hand, in search models, the prediction module is often a similarity calculator, which calculates the similarity score between the query sample and all the database samples, and return the most similar ones as search result.

In traditional content-based product (i.e., product images in online shopping, automatic checkout and other scenarios) search (i.e., find the images of the same category as query image) application, the set of categories in database is often fixed. When new categories with few-shot annotated data are input into the search engine, as argued in [24], the results on both the base categories and new products become inaccurate, while annotating more new-category data or retraining the whole search engine is expensive. This introduces a new problem setting: few-shot incremental product search (FIPS), where new few-shot categories are gradually added to the database of multi-shot base categories, and a query can belong to both the base and new categories.

MetaFA [24] provides a meta-learning solution for this FIPS problem by meta-learning a feature adaptor (FA) across different categories. Technically, the feature adaptor consists of a mean feature adaptor and an attention feature adaptor to discover both the common features and differences among categories. The meta-training for the feature adaptor is based on the series of incremental batches of novel-category data. In this way, the meta-learned feature adaptor could adapt the base feature extractor (pretrained on initial database of base categories) to novel feature extractor, which could extract the visual features of the novel categories. A weight combiner is also designed to gradually update the moving average weights of novel feature extractor. The final prediction for a query is determined as follows: first, calculate the maximum classification score s_x of query image x for the base categories. Next, when $s_x > \tau$ (τ is a predefined threshold), we judge that x belongs to the base categories and use the base feature extractor and search (by similarity) in the base-category database. Otherwise, x belongs to new categories and we use the novel feature extractor and search in the novel-category database. For more details we refer the readers to the original paper [24].

In a nutshell, from a meta-learning perspective, learning the mapping from {base feature extractor weights, few-shot novel category data} to {novel feature extractor weights} is regarded as a task. The goal is to meta-learn this mapping through a series of incremental tasks, and this setting is similar to traditional few-shot learning settings.

Summary

In Sect. 4.1, we have provided an introduction to applying meta-learning to search and recommendation tasks. Sections 4.1.1 and 4.1.3 leverage meta-learning to alleviate the cold-start (i.e. few-shot) issues in recommendation and search applications, to improve the practicability of the recommender systems and search engines on real-world data. The cold-start subjects include users, items, categories and scenarios with few historical data records. In terms of meta-learning approaches, bilevel optimization is the most popular and flexible method which learns across tasks the *hyper-parameters of the learning algorithm* for search and recommender models, including initialization, update policy, stop policy, etc. Memory-based (e.g., MANN [20]) and metric-based (e.g., Siamese Network [10]) meta-learning methods have also been adopted for cold-start recommendation [4, 12] and still need more exploitation on their potential power.

At a higher level beyond hyper-parameter learning, Sect. 4.1.2 takes meta-learning as a tool for learning the *algorithm selection strategy* according to a specific recommendation dataset (or individual user). Traditional methods manually design meta-features for datasets and model the algorithm selection as a classification (for best algorithm) or regression (for performance estimation) problem. In recent work [15], an end-to-end bilevel optimization-based meta-learning strategy is also proposed to automatically learn the selection strategy with implicit meta-features.

Generally speaking, all the methods in this section learn a "prior" knowledge across tasks to inspire more clever search or recommendation in future tasks, which conforms to the spirit of meta-learning and AutoML. In Sects. 3.1 and 3.4, we also manage to automatically find the hyper-parameters (we regard the neural architecture as a special form of algorithm hyper-parameter) of search engines and recommender systems according to the specific data. The main differences among these three sections is the methodology, while they share the common goal: automatic machine learning on new search or recommendation tasks.

As the methods above are designed for general search and recommendation, although to the best of our knowledge, no previous work has adopted meta-learning for *multimedia* search and recommendation, it would be not challenging to apply those methods to multimedia data. As argued in Sect. 3.4, the core stages for both multimedia search and recommendation algorithms is feature extraction and modality fusion. Therefore, as a valuable future direction, it would be effective to apply the above meta-learning methods to the hyper-parameter learning of the two stages of the algorithms. In addition, meta-learning methods for algorithm selection could also be exploited to provide better multimedia recommendation performance on different datasets.

4.2 Meta-Learning for Computer Vision

Visual and textual (language) information are the two most common modalities of data in our daily lives, and there are many essential problems and practical applications on them, such as image classification, object detection, visual captioning, visual tracking, visual question answering etc. Meanwhile, researchers also tried to introduce the meta-learning methodologies to help solve these vision and language tasks, so as to improve the model ability and provide more insights. In this section, we introduce some recent related works as follows.

4.2.1 Meta-Learning for Classification in Images and Videos

Image classification is the most essential problem in visual domain, but current machine learning methods are often hampered when the dataset is small. Recently, Few-Shot Learning (FSL) is proposed to tackle this problem. Using prior knowledge, FSL can rapidly generalize to new tasks containing only a few samples with supervised information [48]. To solve the FSL problem, meta-learning approaches typical involve a meta-learner model that given a few training examples of a new task, it tries to quickly learn a learner model that solves this new task [27, 30, 39, 45, 46].

Specifically, Ravi and Larochelle [43] propose an LSTM-based meta-learner model to learn the exact optimization algorithm used to train another learner neural network classifier in the few-shot regime. The parametrization of their model allows it to learn appropriate parameter updates specifically for the scenario where a set amount of updates will be made, while also learning a general initialization of the learner (classifier) network that allows for quick convergence of training. Mishra et al. [38] propose a class of simple and generic meta-learner architectures that use a novel combination of temporal convolutions and soft attention, with the former aggregating information from past experience and the latter pinpointing specific pieces of information. Qiao et al. [42] propose a novel method that can adapt a pre-trained neural network to novel categories by directly predicting the parameters from the activations. Zero training is required in adaptation to novel categories, and fast inference is realized by a single forward pass. Gidaris et al. [30] enhance a typical object recognition system with an extra component, called few-shot classification weight generator that accepts as input a few training examples of a novel category (e.g., no more than five examples) and, based on them, generates a classification weight vector for that novel category. Khodadadeh et al. [35] propose UMTRA, an algorithm that performs unsupervised, model-agnostic meta-learning for classification tasks. The meta-learning step of UMTRA is performed on a flat collection of unlabeled images. While these images are assumed can be grouped into a diverse set of classes and are relevant to the target task, and no explicit information about the classes or any labels are needed. UMTRA uses random sampling and augmentation to create synthetic training tasks for meta-learning phase. Labels are only needed at the final target task learning step, and they can be as little as one sample per class.

4.2.2 Meta-Learning for Detection in Images and Videos

With the recent advances in deep convolutional neural networks, significant improvements have been achieved in visual detection problems such as object detection in images [31, 33, 44]. However, training effective detectors in a small data regime remains an open challenge. Therefore, detecting objects in a few-shot manner, i.e., learning an object detector for novel classes when only a few examples with annotated bounding boxes are available, is very crucial and has aroused the interest of researchers recent years.

Wu et al. propose a meta-learning framework for object detection named "Meta-RCNN", which learns the ability to perform few-shot detection via meta-learning. Specifically, Meta-RCNN learns an object detector in an episodic learning paradigm on the (meta) training data. This learning strategy helps acquire a prior which enables Meta-RCNN to do few-shot detection on novel tasks. Meta-RCNN is built on the Fast-RCNN framework, with both the Region Proposal Network (RPN) and the object classification branch are meta-learned. The meta-trained RPN learns to provide class-specific proposals, while the object classifier learns to do few-shot classification [50]. Wang et al. [49] apply the meta-learning methodologies to detect rare objects in images. Specifically, they designed a model "MetaDet" built upon the widely-used detection approach Faster R-CNN. The convolutional features and RPN are as category-agnostic components, whose parameters are shared by base and novel classes. While the classifiers and bounding box regressors are category-specific, and the dynamics of their parameters are shared by base and novel classes. Meanwhile, the introduced parameterized weight prediction meta-model is trained on the space of model parameters to predict a category's large-sample bounding box detection parameters from its few-shot parameters. The meta-model is trained end-to-end with Faster R-CNN through a meta-learning procedure. Joseph et al. [34] study the class-incremental object detection problem, where new classes are sequentially introduced to the detection model. To solve this problem, they propose a gradient-based meta-learning approach which learns to reshape the gradients such that optimal updates are achieved for both the old and new tasks. This is efficiently realized by meta-learning a set of parameterized preconditioning matrices, interleaved between layers of standard object detector.

4.2.3   Meta-Learning for Image Captioning

Image captioning [26, 51, 52], which aims to generate sentence descriptions for images, has emerged as a prominent research problem to bridge computer vision and natural language processing. Most image captioning algorithms focus on the supervised setting, which relies on a large amount of human annotated image-text pairs for training. However, in the few-shot image captioning setting, some words are uncommon or even unseen in the training procedure, making these supervised algorithms cannot well handle such novel words. Dong et al. [29] propose Fast Parameter Adaptation for Image-Text Modeling (FPAIT) that learns to learn jointly understanding image and text data by a few examples. On the one hand, FPAIT learns proper initial parameters for the joint image-text learner from a large number of different tasks. On the other hand, FPAIT leverages dynamic linear transformations to alleviate the side effects of the small training set. In this way, FPAIT flexibly normalizes the features and thus reduces the biases during training. Quantitatively, FPAIT achieves superior performance on both few-shot image captioning and VQA benchmarks. In addition, Li et al. [37] propose to use a meta learning methodology to ensure the propositional correctness and distinctiveness of the generated captions whilst optimizing the evaluation metrics. Specifically, a meta model is built to maximize the probability of the ground truth caption as well as maximize the reward of the generated caption. In this approach, different gradient steps are taken to learn these two tasks, simultaneously, which enables the meta model to adapt to the global optimal solution of each task. The optimization of the reward function is thus guided to avoid reward hacking to some extent, and thus ensures the propositional content and distinctiveness of the generated captions.

4.2.4   Meta-Learning for Tracking in Images and Videos

Visual object tracking [32, 36, 53] is a task that locates target objects precisely over a sequence of image frames given a target bounding box at the initial frame. The tracking problem is closely related to the detection problem, and it even can be treated as a special type of object detection, which is called instance detection. Conventional object detection [32, 36, 40, 53] focuses on locating objects of some predefined classes, however, object tracking only looks for a particular instance, which may belong to any known or unknown object class, that is specified in the initial frame. To solve the above problem, Wang et al. [47] try to directly convert a modern object detector into a high-performance tracker, but the main challenge is how to obtain a good initialization of the detector so that once a new instance is given, it can efficiently infuse the instance information into the network without overfitting. Concretely, based on MAML, Wang et al. firstly pick any detector which is trained with gradient descent. Second, they used MAML to train the detector on a large number of tracking sequences. Third, when the initial frame of a test sequence is given, they fine-tune the detector with a few steps of gradient descent. A decent tracker can be obtained after this domain adaptation step. During tracking, when new appearances of the target are collected, the detector can be trained with more samples to achieve an even better adaptation capability.

Besides the above work, Park et al. [41] propose an offline meta-learning-based method to adjust the initial deep networks used in online adaptation-based tracking. The meta learning is driven by the goal of deep networks that can quickly be adapted to robustly model a particular target in future frames. Ideally the resulting models focus on features that are useful for future frames, and avoid overfitting to background clutter, small parts of the target, or noise. By enforcing a small number of update iterations during meta-learning, the resulting networks train significantly faster. Choi et al. [28] propose a novel online visual tracking framework by building a visual tracking system incorporating a Siamese matching network for target search and a meta-learner network for adaptive feature space update. Specifically, they use a fully convolutional Siamese network structure for searching the target object in a given frame, where target search can be done fast and efficiently using the cross-correlation operations between feature maps. For the meta-learner network, they propose a parameter prediction network inspired by recent advances in the meta learning methodology for few-shot learning problems. By incorporating the meta-learner network, the target-specific feature space can be constructed instantly with a single forward pass without any iterative computation for optimization and is free from innate overfitting, improving the performance of the tracking algorithm.

4.2.5   Meta-Learning for Visual Question Answering

We would like to point out that, as an emerging research topic that touches both visual and textual information, visual question answering (VQA) has few interactions with meta-learning so far. Therefore, in this section we first present some basic concepts including some taxonomies and datasets for VQA and discuss several works that are most relevant to the combination of meta-learning and visual question answering as well as potential future directions that may push the research.

Visual question answering (VQA) has gradually attracted the interests of researchers these years, for its nature that combines vision and language domain, which are the two most active and vigorous field in machine learning recently. The difference from "Captioning" mentioned earlier is, VQA include additional textual input, the "question", to specify the region or relation of interests that the agent should attend to. In a simplified way, this problem can be formulated as follow:

$$\tilde{T}^A = M(I, T^Q), \tag{4.1}$$

where T^A, M, T^Q denote the text of answers, the model, and the text of question respectively. This creates needs for sequence processing and multi-modal alignment abilities for proposed methods, not only at the decoding stage, but also for the encoding and parsing for the input. In form, it is more complicated compared with those vision-to-language-mapping tasks. The model M needs to understand the question, referring to the images or videos, and find the answer. More cross-modal interactions are required.

The contents for VQA tasks vary a lot, where contents here refer to both vision and language part. In temporal dimension, the visual input could be static image, or short clips, or long-term videos. And for visual contents, it could be natural images or synthetic scenes, etc. For a good definition for natural image here, we can refer to the Deep Learning book by Goodfellow et al. [63]:

> A natural image is an image that might be captured by a camera in a reasonably ordinary environment, as opposed to a synthetically rendered image, a screenshot of a web page, etc.

And for each temporal and visual contents setting, there is a language/question field corresponding to it. Hence the questions and answers also differ a lot in temporal type (descriptive or predictive), goal of interests (query for attributes or query for a reason), causal type (explanation or counterfactual) and so on. In practice, a specific VQA task is defined directly by a dataset, including both visual input and Q&A pairs. Note that it's quite common that several VQA datasets share visual inputs while differ in text annotations. For example, the CLEVR-Humans dataset [69] and original CLEVR dataset [68]. Figure 4.8 shows their differences. We will introduce more datasets later in detail.

We will first introduce the Visual Question Answering (VQA) tasks, along with the corresponding datasets, classic papers and methods. Then we will turn to the combination of meta-learning and VQA, and introduce some articles in this area. Finally we give a brief summary of this section.

A. Visual QA Tasks and Datasets

A large number of VQA datasets have emerged in recent years. Due to space limitations, we choose some popular ones to introduce, then we will give a rough classification according to the task at the end of this part.

I. VQA on Static Images

MS COCO (Microsoft Common Objects in Context) [77]   Strictly speaking, this is not a VQA dataset, for the absence of Q&A text annotations. However, as mentioned before, its visual contents provides the foundation for many VQA datasets, such as Visual7W [101], Visual Genome [71] and VisDial [59]. This dataset is originally collected for large-scale object detection, segmentation, key-point detection and captioning, including 328K images, and has rich annotation labels for these tasks. Most of these tags may be used as an aid to VQA issues. Figure 4.4 shows an example from this dataset.

Captions	Labels:
a green passenger bus is boarding passengers near some water.	person
a large bus is parked outside by a river	bus
a green bus that is sitting on a street.	bench
bus moving past small pond with people sitting on benches next to it	backpack
a green bus is parked near some water	handbag

Fig. 4.4 Examples from MS-COCO dataset

Question Answers		Relationship
What color is the cat?	White and orange.	cat has ears
How many eyes does the cat have?	Two.	Region
Where is the dresser?	Right side of picture.	orange and pink cat ears
What is on the cats face?	Whiskers.	Attribute
What is the cat doing?	Relaxing on bed.	ears is orange

Fig. 4.5 Examples from visual genome dataset

Visual Genome/Visual7W [71, 101] As mentioned before, this two datasets both share a part of visual contents with MS-COCO. Among them, the scale of Visual Genome is larger, which consists of 101k images with 1.7 million QA pairs. The questions follow balanced distributions over 6 types: What, Where, When, Why, Who and How. Visual7W has similar settings but one more "W", i.e., Which, this is why it's called seven Ws. It contains 47K images with 0.3 million QA pairs. Besides multiple choices questions, "which" leads to a new type of Q&A, object grounding. Figure 4.5 shows the examples.

Question : What animals are in this picture?

Fig. 4.6 Examples from VQA dataset

Visual Question Answering (VQA) Dataset [64] Different from most similar datasets, questions on images from this dataset is open-ended, makes higher requirements for the output part of the model, not only the ability for classification from finite choices, but also the ability for sentence generation. This is also an early classic dataset, Fig. 4.6 shows some visual examples.

II. VQA on Video Clips

TGIF-QA [67] The previous datasets focus on single-frame pictures, this TGIF-QA makes a step forward to QA on long video sequences. GIF is a convenient short-term video format, which can be seen as a transition between single-frame images and long-term videos. TGIF-QA shares visual contents with TGIF [76], and recollect 165K QA pairs via crowdsourcing. There are some questions obviously reflects the necessity of video time domain information when answering questions. Figure 4.7 shows some examples.

MSVD and MSR-VTT [56, 96] The Microsoft Research Video Description Corpus (MSVD) including more than 2K video snippets and about 120K sentences. The original text annotations are action summaries, while later workers used some tools to automatically convert the text summarization to QA pairs. The situation is similar for Microsoft Research Video to Text (MSR-VTT), which has a larger scale, with 10K video clips and 20 sentences for each video clip. The text to QA conversion can be found in TGIF-QA paper [67].

TVQA and TVQA+ [72, 73] These two datasets are based on six popular TV shows (Friends, The Big Bang Theory, How I Met Your Mother, House M.D., Grey's Anatomy, Castle), including 21.8K clips and 152.5K QA pairs. One unique feature of this task is that there exist subtitles as additional text

Caption: a kitten climbs a bunk bed ladder to the top bunk.
Q: What does the cat do 4 times ?
a) bob head around
b) move upward with rod
c) step
d) turn
e) workout on bar
Ans: b)

Fig. 4.7 Examples from TGIF-QA dataset

inputs, which requires the model be able to comprehensively consider the input information of multiple modalities. TVQA+ relabels the precise event boundaries to help temporal grounding.

III. Visual Dialog

VisDial [59] Dialogue is a generalized version of single-round QA, contextual long-term dependencies need to be considered in this situation. Visual Dialog (VisDial) dataset also shares visual contents with MS COCO dataset, and contains 123K/ 2K/ 8K dialogues for training/ validation/ test split in the version 1.0. For each image, there are two annotator, playing the role of questioner and answerer, chatting about the image to help the questioner better imagine the scene.

IV. VQA with Reasoning

CLEVR and CLEVR-Humans [68, 69] Compositional Language and Elementary Visual Reasoning (CLEVR) form a synthetic dataset, containing 3D rendered objects with several attributes and including size, shape, material as well as color. The questions are a little more complicated compared to classic VQA questions, requiring inference using relations, comparison, or counting. Besides QA pairs, there are also scene graph and functional program annotations, which help to show the reasoning path. The CLEVR-Humans

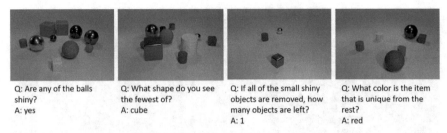

Q: Are any of the balls
shiny?
A: yes

Q: What shape do you see
the fewest of?
A: cube

Q: If all of the small shiny
objects are removed, how
many objects are left?
A: 1

Q: What color is the item
that is unique from the
rest?
A: red

Fig. 4.8 Examples from CLEVR-Humans dataset, with unrestricted words and phrases annotations

dataset shares the visual contents, and recollects human-generated natural questions where there are more out-of-vocabulary words, requiring the ability to deal with uncommon words. Figure 4.8 shows some examples.

CLEVRER and CATER [62, 98] Collision Events for Video Representation and Reasoning (CLEVRER) form a synthetic video dataset, containing moving 3D objects and collision events. It includes four types of questions, descriptive (e.g., "what color"), explanatory ("what is responsible for"), predictive ("what will happen next"), and counterfactual ("what if"). These settings introduce new reasoning tasks and requirements, widening the boundary of visual reasoning problems. Compositional Actions and Temporal Reasoning (CATER) dataset is a similar synthetic dataset, where the motion part is more complex, including rotation and overlapping, which demonstrates the exploration of the complex relationship between moving objects in the nature.

VCR [100] Visual Commonsense Reasoning (VCR) dataset contains 110K movie scenes, and has 212K/26K/25K questions as training/validation/test sets. Besides answering the main question, this task requires to answer a sub-question explaining why a particular answer is chosen. Formally, it acts as a two-stage VQA task, while the relevance between the two questions makes it helpful for obtaining a clearer reasoning path. Figure 4.9 shows some examples.

V. Other VQA Tasks and Datasets

TextVQA [86] This is a dataset designed for reading and understanding texts in images. Roughly speaking, the optical character recognition (OCR) function should be embodied in the model to read texts in images first. It's a new task, and the model need the ability to handle inputs from multiple modalities.

GQA [66] GQA is a VQA dataset which contains real natural images, sharing visual contents with Visual Genome, while with high-quality scene graphs on object classes, attributes, and their relations. It is suitable for reasoning on real images, and has been well studied these years. There are also classification and object detection features that serve as auxiliary aids.

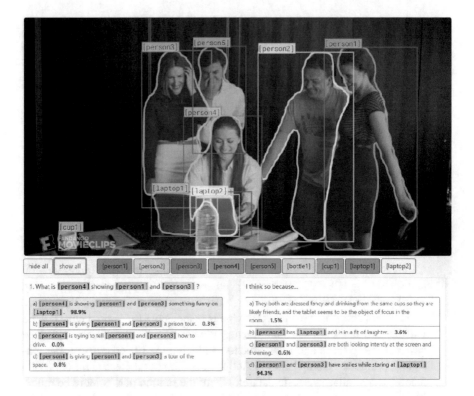

Fig. 4.9 Examples from VCR dataset

Each of the above datasets defines a task, or a benchmark with corresponding applicable methods on these benchmarks, respectively. We will introduce them according to different categories in the following.

B. Classic Methods for VQA Tasks

Wu et al. write a good review [92] to discuss the state-of-the-art methods, dividing these methods into four main categories:

I. Joint Embedding Approaches

The motivation comes from the development of natural language processing. Comparing to captioning, VQA has an additional inference stage among visual and linguistic modalities. Normally, image representations come from pretrained CNN, text representations come from pre-trained word embedding. Malinowski et al. [81] propose "Neural-Image-QA" model, using LSTM+RNN to deal with text information, and pretrained CNN to deal with image features, then feed the two features into LSTM encoder and LSTM decoder, to generate sentences with variable lengths, until <end> is generated. For this type of method, if the task has multiple choices instead of open-ended questions, just replace the LSTM encoder and decoder with classifier layers. Method proposed by Ren et al. [84] is a variation of this method. Gao et al. [61] propose "Multimodel QA" model, using LSTM to encoding questions and generate answers. Note here LSTM encoder and decoder share weights. Noh et al. [83] utilize "DPPNet" to extract image features, use GRU units to deal with question features, and adopt fully-connected layers as classification layers to generate class weights. Fukui et al. [60] propose a pooling-based method to embed features, named "Multiple Compact Bilinear pooling" (MCB). Kim et al. [70] resort to MRN (multimodal residual learning framework) for learning joint embedding. Saito et al. [85] propose DualNet for fusing operation, performing addition or multiplication among corresponding elements of each kind of features. Ma et al. [80] employ CNN to extract both image and text features and instead utilize RNN for text features before feeding them into multimodal CNN for joint embedding. This type of methods (joint embedding) are straight forward, serving as the basis of most VQA methods, though there are still a lot of room for improvement.

II. Attention Mechanisms

The aforementioned joint embedding methods use global information as input, this may introduce extraneous noise. Attention mechanisms are proposed to accurately locate the region of interests, making it easier for the later part to analyze information and output answers. Zhu et al. [101] introduce how to combine attention mechanisms with LSTM, while Chen et al. [57] use question-guided attention map for searching relevant parts in the image corresponding to question semantics, and Yang et al. [97] use stacked attention networks (SAN) to generate answers iteratively. Xu et al. [95] propose multi-hop image attention scheme (SMem), using words and questions to guide attentions. Lu et al. [79] propose hierarchical co-attention model (HieCoAtt) through combining visual attentions. Fukui et al. [60] add the attention mechanism to multimodal compact bilinear pooling (MCB). Attention mechanism is able to improve the model performance which uses global features as input previously. However, the attention module includes more parameters, which needs to be carefully designed to avoid parameter redundancy and reduced parameter efficiency.

III. Models with External Knowledge Bases

The motivation is that, when performing visual question answer tasks, some non-visual prior information is often needed, including commonsense or expert knowledge. There are some existing knowledge bases, such as DBpedia, Freebase, YAGO, OpenIE, NELL, WebChild, ConceptNet etc. Wang et al. [91] propose DBpedia based VQA model named "Ahab". The model extracts visual concepts first, then combines similar concepts in DBpedia, and feeds to a learnable process mapping image-question to queries, followed by summarizing query results for answers. FVQA proposed by Wu et al. [93] is an improved model based on this method. This work proposes a joint embedding using external knowledge, it retrieves external knowledge of relevant attributes from DBpedia, and embeds them into the word vector with Doc2Vec. The other part is similar to the joint embedding framework, feeding CNN visual features and text embeddings into LSTM, to explain the questions and generate answers. This type of methods are effective, although in restricted question types. Since the publication of this review, there have been many new developments in the research of VQA these years.

IV. **Visual and Textual Transformer**

The success of transformer in the field of NLP has also spread to VQA. There are a lot of works focusing on utilizing transformer to generate vision and language joint embeddings [54, 74, 75, 78, 87–89]. Sun et al. [88] concatenate video features directly after the text embeddings, then feed them together into transformer to generate vision and language joint embeddings for follow-up operations. Tan et al. [89] construct a similar transformer architecture, and pre-trained the model with a large number of (image, sentence) pairs, making the model be able to connect visual and language semantics. Five different and representative pre-training tasks were used, including masked language modeling, masked object prediction (feature regression and label detection), cross-modal alignment, and image QA. Note that they aim to produce a generalized joint embedding method, image QA is only one of the pre-training tasks. These methods using vision and language transformer can be seen as a combination of the above two methods, joint embedding and attention based methods, because the purpose of transformer here is to obtain joint embedding, with the part of Q, K, V vectors in transformer embed attention mechanism. The advantage is the better performances and generalized joint representations. However, the inner processing in the transformer is invisible, making it difficult to interpret. On the other hand, the transformer module also includes plenty of parameters, especially for the visual features. The balance between the performance and parameter efficiency becomes a trade-off.

V. **Compositional Models**

This type of method has also been updated in recent years. We can roughly divide it into two categories, i.e., implicit and explicit.

"MAC" proposed by Hudson et al. [65] is a representative implicit method which employs recurrent cells to decompose the original questions. The attention mask of each step in MAC is visible, making it possible to supervise the reasoning process. In each cell, the question guides the reading process from visual base, and directs the generation of output to next cell along with the read contents and memory from the last cell. Two flows of control and memory transfer through cells, and an additional classifier accepting question and the output of final cell predicts the answer. This work makes a step forward upon the interpretability of the reasoning process. Mid-term attention masks validate that the model can accurately discover every related object, proving the ability to understand phrases describing object and attribute relationships. The implicit manner and learnable recurrent modules make it generalizable to other similar tasks and datasets. The only disadvantage is that the visualization of reasoning process is not as clear as the methods which to be discussed in the following.

Andreas et al. [55] propose neural module networks (NMN) which link and combine different modules for better supervision during the reasoning process. Their contributions lie in the use of logical reasoning for continuous visual features instead of discrete or logical prediction. Besides images and labels, the method shows attention maps related to some key words during the reasoning process. This makes the model better at reasoning than traditional models, capable of dealing with more complicated questions. As an early work, there exist several flaws at the question parsing stage—the authors simplify questions so much that some information may be missed. Xiong et al. [94] apply dynamic memory networks (DMN) to VQA, where the model consists of four independent modules, (i) input module vertorizing input data, (ii) question module using GRU to get text representations, (iii) episodic memory module retrieving for the answers, and (iv) answer module producing outputs. Compared with NMN, this method performs similar in two-value problems while a little worse in mathematical problems, and performs better in other problems.

Recently, there is another branch that is becoming very popular, i.e., the neural symbolic methods, which explain the reasoning process in an explicit manner, using a list of short-term "programs" to describe the reasoning path. Yi et al. [99] propose a NS-VQA model, which consists of three parts, (i) neural scene parsing module processing input images through Mask R-CNN and CNN feature extractor to generate abstract scene representations including object IDs, attributes and positions, (ii) neural question parsing module matching input questions to programs using LSTM encoder, and (iii) symbolic program execution module accepting output from these two modules and executing the programs to generate answers. NS-DR proposed by Yi et al. [98] is a generalized video version of NS-VQA. Besides the video parser, question parser and program executor, there is another module named dynamics predictor, which is specially designed for temporal prediction in videos. This module accepts scene representation from video parser, and predicts the future position of each object. This model is also able to solve predictive and counterfactual problems. Mao et al. [82] propose an improved version, i.e., the NS-CL model, where the symbolic reasoner is slightly different, and the visual scene representation as well as the text concept embeddings are designed to participate in the calculation instead of performing pre-defined fixed program executors. In addition, the neural symbolic reasoning module utilizes reinforcement in the program parser, helping to finish language parsing even without learning. This method shows excellent data efficiency, compared to MAC. The neural symbolic methods show explicit reasoning path and great data efficiency. However, they are restricted by the program set and hand-written executor, making it difficult to generalize to real-world scenes and complex language situations.

VI. Other Methods

There are some other methods in which the external knowledge base is extended for the expansion of QA contents, and new knowledge bases are developed etc. The research works on reasoning tend to focus more on those sophisticated reasoning problems, such as predictive and counterfactual cases. Video QA with temporal dependencies also attracts some attentions. And there also exist some new tasks and datasets related to VQA, including querying on physical motions that have not happened yet, or asking for trajectories in videos etc.

C. Marrying Meta-Learning with VQA

In this section, we discuss a few existing works relevant to the idea of "meta-learning for VQA", and point out possible ways for further combinations of meta-learning and VQA in order to provide some inspirations to the readers.

Chen et al. [58] propose a meta module network (MMN) to combine advantages of both explicit multi-hop neural module network and monolithic network with implicit reasoning in the latent feature space. The former excels in interpretability and compositionality, while the latter usually achieves better performance due to model flexibility and parameter efficiency. They utilize a meta module to generate candidate recipes as input for substantiation of the designed instance modules to accomplish different sub-tasks. This work is similar to neural symbolic methods, and the instance modules correspond to programs. The meta module is named "meta" because it learns to learn instance operations.

Teney et al. [90] introduce a meta learning problem setting, the model is trained first on a small training set, and is provided with a large additional "support set" at testing time. Then the model must learn to learn, or to exploit additional data on-the-fly, without retraining. This setting is more similar to meta-learning, or continuous learning, which can indeed improve the generalization performance and environmental adaptability of the model. Moreover, the authors introduce many dynamic parameters which are determined by the memory and support set. The final output is produced using both static and dynamic weights which is a good attempt to "meta VQA models".

We also find that the two "meta + " work introduced above possess different "meta" parts. The first meta module network focuses on model and learning process, aiming to find a high level module to direct low level instance module which performs simple VQA tasks. From this aspect, high level models of the compositional branch can be seen as meta modules to some extent. The second work focuses on source data input and problem definition, aiming to improve the generalization and continuous learning ability of VQA models.

We can give an outline of the future "meta + VQA" work based on these two categories. One direction is model-based, which extends program sets or form a more complex execution network. The challenge lies in acquiring supervision for intermediate outputs, or designing an automatic mechanism to verify their rationality. At a glance, reinforcement learning seems to be useful. The other direction is data-based or learning-based, which uses as less data samples as possible, and keeps learning new concepts with high efficiency. The challenge may lie on how to achieve high data efficiency while maintaining considerable performances, as well as how to manage memories for better trade-off over previous and current performances.

Summary
In this section, we introduce visual question answering (VQA) tasks, give taxonomies, as well as present several classic datasets. Then, we briefly discuss corresponding methods on these benchmarks including both classic papers and recent trends. We also discuss two works applying meta-learning to VQA tasks, and also share our insights on future work in each direction. Although the VQA tasks have been well-developed in the past decade, applying meta-learning to VQA still remains largely unexplored. There is plenty of room left for exploring the marriage between meta-learning and VQA, which offers great opportunities for further investigations.

4.3 Meta-Learning for Natural Language Processing

Meta-learning has achieved great success to be employed in quantities of artificial intelligence areas, especially in computer vision. There still remain a lot of spaces to explore by applying meta-learning to the natural language processing (i.e., NLP) domain, especially under the situation when tasks only have limited labeled training examples (i.e., few-shot NLP). With the help of meta-learning, which is to provide effective parameter initialization after training on a variety of other tasks with rich annotations, those new tasks with fewer labeled examples can perform pretty well after only a couple of gradient updates.

In this section, we will give a brief introduction to demonstrate the current research progress of employing meta-learning methods, the typical optimization-based meta-learning method MAML [106], in many few-shot NLP applications (e.g., few-shot text classification, low-resource neural machine translation, and so on). Additionally, we summarize existing datasets for few-shot NLP applications and point out some shortcomings they have as well.

4.3.1 Meta-Learning for NLP Classification Tasks

As few-shot learning serves as one of the popular scenarios that meta-learning has been widely adopted, we discuss NLP classification in terms of two tasks: few-shot text classification and few-shot relation classification.

A. Few-Shot Text Classification

Yin et al. [124] distinguish few-shot text classification task into two categories. One is to view each class as a task for multi-class classification problems. The other one is to view each domain as a task with the same classification problem.

Jiang et al. [112] propose a model belonging to the former, which introduces Attentive Task-Agnostic Meta-Learning (ATAML) algorithm inspired by MAML [106] for text classification. And others [108, 125] belong to the latter, both handling the multi-domain problems from two datasets: one is a multi-domain sentiment classification dataset and the other is a multi-domain real-world dialogue intent dataset. Specifically, Yu et al. [125] propose an adaptive metric learning approach that automatically determines the best-weighted combination from a set of metrics obtained from meta-training tasks for a newly seen few-shot task. Geng et al. [108] present a novel Induction Network to learn such a generalized class-wise representation, which can induce and generalize better by innovatively leveraging the dynamic routing algorithm in meta-learning.

B. Few-Shot Relation Classification

Relation classification (RC) tasks are suffering from unavailable large-scale training datasets, which thus primarily rely on distant supervision (DS). However, it can also be viewed as a few-shot problem.

Gao et al. [107] formalize RC task as a few-shot problem and proposed hybrid attention-based prototypical networks to solve such a noise few-shot problem due to the diversity and noise of text. Obamuyide et al. [118] present a model-agnostic meta-learning protocol for training relation classifiers to achieve enhanced predictive performance in limited supervision settings.

4.3.2 Multi-Domain Low-Resource Natural Language Generation

We discuss multi-domain low-resource natural language generation with respect to tasks and datasets. In particular, the tasks include few-shot dialogue system, neural machine translation, knowledge graph completion etc.

A. Few-Shot Dialogue Systems

Mi et al. [117] formulate the natural language generation (NLG) problem under low-source setting from a meta-learning perspective, and proposed a generalized optimization-based approach (Meta-NLG) based on the well-recognized model-agnostic meta-learning (MAML) algorithm, which defines a set of meta tasks, and directly incorporates the objective of adapting to new low-resource NLG tasks into the meta-learning optimization process. Madotto et al. [116] also extend MAML to learn quickly adapting to new personas by leveraging only a few dialogue samples collected from the same user, which is fundamentally different from conditioning the response on the persona descriptions. DAML [119] learns across multiple rich-resource tasks by applying MAML to the dialog domain so that it can efficiently adapt to new domains with minimal training instances. To conclude, all these methods do not change the original MAML, they directly apply it to their scenarios due to the model-agnostic property of MAML.

B. Low-Source Neural Machine Translation

MetaNMT [109] is proposed to extend MAML for low-resource neural machine translation (NMT), framing low-resource translation as a meta-learning problem. Figure 4.10 shows the training process of proposed MetaNMT. For each episode, one language pair is sampled as a task for meta-learning. MetaNMT learns to adapt to low-resource languages (i.e., five diverse languages) based on multilingual high-resource language tasks (i.e., eighteen European languages).

4.3.3 Knowledge Graph (KG) Completion

MetaR [103] is proposed to solve the task of few-shot link prediction in KGs (i.e., to predict new triples about a relation by only observing a few associative triples), by focusing on transferring relation-specific meta information to make the model learn the most important knowledge and learn faster.

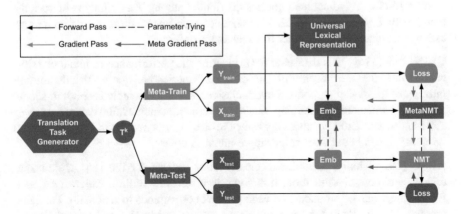

Fig. 4.10 Training process of MetaNMT

And Meta-KGR [115] is another effective and explainable method for multi-hop knowledge graph reasoning, which adopts meta-learning to learn effective meta parameters from high-frequency relations that could quickly adapt to few-shot relations.

4.3.4   Other Tasks

There are also many other methods aiming to solve specific NLP tasks with meta-learning. Holla et al. [111] propose a meta-learning framework for few-shot word sense disambiguation (WSD), where the goal is to learn to disambiguate unseen words from only a few labeled instances. Others [105, 113] reformulate the event detection (ED) task with limited labeled data as a few-shot learning problem.

4.3.5   Datasets for Few-Shot NLP Tasks

There are quite a lot of benchmark datasets for few-shot NLP tasks, pushing the research progress of this area, even though some of them only consider the within-dataset generalization. The detailed descriptions for each of them are listed as follows.

FewRel [110] This is a relation classification dataset with 100 relations, each of which has 700 labeled sentences. Usually, it is split into 64/16/20 relations as training/validation/testing sets. This dataset is manually annotated from Wikipedia corpus by crowdworkers. The relation of each sentence is first recognized by distant supervision methods and then filtered by crowdworkers.

SNIPS [104] The SNIPS Natural Language Understanding benchmark is a dataset of over 16,000 crowd-sourced queries distributed among 7 user intents of various complexity. Several works [122, 123] conduct experiments on it using two intents as few-shot classes and the other five intents for training.

CLINC150 [114] This dataset is for evaluating the performance of intent classification systems in the presence of "out-of-scope" queries(i.e., queries that do not fall into any of the system-supported intent classes). The dataset includes both in-scope and out-of-scope data, containing 23,700 instances in which 22,500 examples cover 150 intents, and 1200 instances are out-of-scope. All queries are in English and all samples/queries in our dataset are single-intent samples.

ARSC [102] This is a sentiment classification dataset for the task of Amazon review sentiment classification. It is comprised of a few million Amazon reviews for 23 categories of products and each product corresponds to a domain. For each product domain, there are three different binary classification tasks according to three different thresholds against review ratings. Therefore this dataset contains 69 tasks in total, among which a couple of works [120, 125] use 12 tasks as target tasks and the remaining for training.

However, despite the occurrence of existing benchmark datasets, Triantafillou et al. [121] claim that there should be more datasets that are challenging and consistent with reality. Both FewRel and CLINC150 mainly concentrate on homogeneous learning tasks, ignoring the class imbalance in real life. These models may not be able to generalize to tasks of new distributions.

Summary

We demonstrate the progress of applying meta-learning to the natural language domain in this part. We can see that there do exist quite a number of works that employ meta-learning techniques, especially MAML, to diverse sub-areas of NLP and achieve brilliant performance as well.

References

1. Hospedales, Timothy and Antoniou, Antreas and Micaelli, Paul and Storkey, Amos. Meta-learning in neural networks: A survey. In *arXiv preprint arXiv:2004.05439*, 2020.
2. Homanga Bharadhwaj. Meta-learning for user cold-start recommendation. In *2019 International Joint Conference on Neural Networks (IJCNN)*, pages 1–8. IEEE, 2019.
3. Tiago Cunha, Carlos Soares, and André CPLF de Carvalho. Metalearning and recommender systems: A literature review and empirical study on the algorithm selection problem for collaborative filtering. *Information Sciences*, 423:128–144, 2018.
4. Manqing Dong, Feng Yuan, Lina Yao, Xiwei Xu, and Liming Zhu. Mamo: Memory-augmented meta-optimization for cold-start recommendation. *arXiv preprint arXiv:2007.03183*, 2020.
5. Zhengxiao Du, Xiaowei Wang, Hongxia Yang, Jingren Zhou, and Jie Tang. Sequential scenario-specific meta learner for online recommendation. In *Proceedings of the 25th ACM SIGKDD International Conference on Knowledge Discovery & Data Mining*, pages 2895–2904, 2019.
6. Alex Graves, Greg Wayne, and Ivo Danihelka. Neural turing machines. *arXiv preprint arXiv: 1410.5401*, 2014.
7. Huifeng Guo, Ruiming Tang, Yunming Ye, Zhenguo Li, and Xiuqiang He. Deepfm: a factorization-machine based neural network for ctr prediction. *arXiv preprint arXiv:1703.04247*, 2017.
8. Sepp Hochreiter and Jürgen Schmidhuber. Long short-term memory. *Neural computation*, 9(8):1735–1780, 1997.
9. Yuchin Juan, Yong Zhuang, Wei-Sheng Chin, and Chih-Jen Lin. Field-aware factorization machines for ctr prediction. In *Proceedings of the 10th ACM Conference on Recommender Systems*, pages 43–50, 2016.
10. Gregory Koch, Richard Zemel, and Ruslan Salakhutdinov. Siamese neural networks for one-shot image recognition. In *ICML deep learning workshop*, volume 2. Lille, 2015.
11. Hoyeop Lee, Jinbae Im, Seongwon Jang, Hyunsouk Cho, and Sehee Chung. Melu: Meta-learned user preference estimator for cold-start recommendation. In *Proceedings of the 25th ACM SIGKDD International Conference on Knowledge Discovery & Data Mining*, pages 1073–1082, 2019.
12. Ruirui Li, Xian Wu, Xian Wu, and Wei Wang. Few-shot learning for new user recommendation in location-based social networks. In *Proceedings of The Web Conference 2020*, pages 2472–2478, 2020.

13. Zhenguo Li, Fengwei Zhou, Fei Chen, and Hang Li. Meta-sgd: Learning to learn quickly for few-shot learning. *arXiv preprint arXiv:1707.09835*, 2017.
14. Yuanfu Lu, Yuan Fang, and Chuan Shi. Meta-learning on heterogeneous information networks for cold-start recommendation. 2020.
15. Mi Luo, Fei Chen, Pengxiang Cheng, Zhenhua Dong, Xiuqiang He, Jiashi Feng, and Zhenguo Li. Metaselector: Meta-learning for recommendation with user-level adaptive model selection. In *Proceedings of The Web Conference 2020*, pages 2507–2513, 2020.
16. Nitin Mishra, Vimal Mishra, and Saumya Chaturvedi. Tools and techniques for solving cold start recommendation. In *Proceedings of the 1st International Conference on Internet of Things and Machine Learning*, pages 1–6, 2017.
17. Feiyang Pan, Shuokai Li, Xiang Ao, Pingzhong Tang, and Qing He. Warm up cold-start advertisements: Improving ctr predictions via learning to learn id embeddings. In *Proceedings of the 42nd International ACM SIGIR Conference on Research and Development in Information Retrieval*, pages 695–704, 2019.
18. Steffen Rendle. Factorization machines. In *2010 IEEE International Conference on Data Mining*, pages 995–1000. IEEE, 2010.
19. Andrei A Rusu, Dushyant Rao, Jakub Sygnowski, Oriol Vinyals, Razvan Pascanu, Simon Osindero, and Raia Hadsell. Meta-learning with latent embedding optimization. *arXiv preprint arXiv:1807.05960*, 2018.
20. Adam Santoro, Sergey Bartunov, Matthew Botvinick, Daan Wierstra, and Timothy Lillicrap. Meta-learning with memory-augmented neural networks. In *International conference on machine learning*, pages 1842–1850, 2016.
21. Jake Snell, Kevin Swersky, and Richard Zemel. Prototypical networks for few-shot learning. In *Advances in neural information processing systems*, pages 4077–4087, 2017.
22. Manasi Vartak, Arvind Thiagarajan, Conrado Miranda, Jeshua Bratman, and Hugo Larochelle. A meta-learning perspective on cold-start recommendations for items. In *Advances in neural information processing systems*, pages 6904–6914, 2017.
23. Oriol Vinyals, Charles Blundell, Timothy Lillicrap, Daan Wierstra, et al. Matching networks for one shot learning. In *Advances in neural information processing systems*, pages 3630–3638, 2016.
24. Qi Wang, Xinchen Liu, Wu Liu, An-An Liu, Wenyin Liu, and Tao Mei. Metasearch: Incremental product search via deep meta-learning. *IEEE Transactions on Image Processing*, 29:7549–7564, 2020.
25. Ronald J Williams. Simple statistical gradient-following algorithms for connectionist reinforcement learning. *Machine learning*, 8(3-4):229–256, 1992.
26. Peter Anderson, Xiaodong He, Chris Buehler, Damien Teney, Mark Johnson, Stephen Gould, and Lei Zhang. Bottom-up and top-down attention for image captioning and visual question answering. In *Proceedings of the IEEE conference on computer vision and pattern recognition*, pages 6077–6086, 2018.
27. Marcin Andrychowicz, Misha Denil, Sergio Gomez, Matthew W Hoffman, David Pfau, Tom Schaul, Brendan Shillingford, and Nando De Freitas. Learning to learn by gradient descent by gradient descent. In *Advances in neural information processing systems*, pages 3981–3989, 2016.
28. Janghoon Choi, Junseok Kwon, and Kyoung Mu Lee. Deep meta learning for real-time target-aware visual tracking. In *Proceedings of the IEEE International Conference on Computer Vision*, pages 911–920, 2019.
29. Xuanyi Dong, Linchao Zhu, De Zhang, Yi Yang, and Fei Wu. Fast parameter adaptation for few-shot image captioning and visual question answering. In *Proceedings of the 26th ACM international conference on Multimedia*, pages 54–62, 2018.

30. Spyros Gidaris and Nikos Komodakis. Dynamic few-shot visual learning without forgetting. In *Proceedings of the IEEE Conference on Computer Vision and Pattern Recognition*, pages 4367–4375, 2018.
31. Ross Girshick. Fast r-cnn. In *Proceedings of the IEEE international conference on computer vision*, pages 1440–1448, 2015.
32. Qing Guo, Wei Feng, Ce Zhou, Rui Huang, Liang Wan, and Song Wang. Learning dynamic siamese network for visual object tracking. In *Proceedings of the IEEE international conference on computer vision*, pages 1763–1771, 2017.
33. Kaiming He, Georgia Gkioxari, Piotr Dollár, and Ross Girshick. Mask r-cnn. In *Proceedings of the IEEE international conference on computer vision*, pages 2961–2969, 2017.
34. KJ Joseph, Jathushan Rajasegaran, Salman Khan, Fahad Shahbaz Khan, Vineeth Balasubramanian, and Ling Shao. Incremental object detection via meta-learning. *arXiv preprint arXiv:2003.08798*, 2020.
35. Siavash Khodadadeh, Ladislau Boloni, and Mubarak Shah. Unsupervised meta-learning for few-shot image classification. In *Advances in Neural Information Processing Systems*, pages 10132–10142, 2019.
36. Matej Kristan, Ales Leonardis, Jiri Matas, Michael Felsberg, Roman Pflugfelder, Luka Cehovin Zajc, Tomas Vojir, Gustav Hager, Alan Lukezic, Abdelrahman Eldesokey, et al. The visual object tracking vot2017 challenge results. In *Proceedings of the IEEE international conference on computer vision workshops*, pages 1949–1972, 2017.
37. Nannan Li, Zhenzhong Chen, and Shan Liu. Meta learning for image captioning. In *Proceedings of the AAAI Conference on Artificial Intelligence*, volume 33, pages 8626–8633, 2019.
38. Nikhil Mishra, Mostafa Rohaninejad, Xi Chen, and Pieter Abbeel. A simple neural attentive meta-learner. *arXiv preprint arXiv:1707.03141*, 2017.
39. Tsendsuren Munkhdalai and Hong Yu. Meta networks. *Proceedings of machine learning research*, 70:2554, 2017.
40. Guanghan Ning, Zhi Zhang, Chen Huang, Xiaobo Ren, Haohong Wang, Canhui Cai, and Zhihai He. Spatially supervised recurrent convolutional neural networks for visual object tracking. In *2017 IEEE International Symposium on Circuits and Systems (ISCAS)*, pages 1–4. IEEE, 2017.
41. Eunbyung Park and Alexander C Berg. Meta-tracker: Fast and robust online adaptation for visual object trackers. In *Proceedings of the European Conference on Computer Vision (ECCV)*, pages 569–585, 2018.
42. Siyuan Qiao, Chenxi Liu, Wei Shen, and Alan L Yuille. Few-shot image recognition by predicting parameters from activations. In *Proceedings of the IEEE Conference on Computer Vision and Pattern Recognition*, pages 7229–7238, 2018.
43. Sachin Ravi and Hugo Larochelle. Optimization as a model for few-shot learning. 2016.
44. Shaoqing Ren, Kaiming He, Ross Girshick, and Jian Sun. Faster r-cnn: Towards real-time object detection with region proposal networks. In *Advances in neural information processing systems*, pages 91–99, 2015.
45. Adam Santoro, Sergey Bartunov, Matthew Botvinick, Daan Wierstra, and Timothy Lillicrap. One-shot learning with memory-augmented neural networks. *arXiv preprint arXiv:1605.06065*, 2016.
46. Sebastian Thrun. Lifelong learning algorithms. In *Learning to learn*, pages 181–209. Springer, 1998.
47. Guangting Wang, Chong Luo, Xiaoyan Sun, Zhiwei Xiong, and Wenjun Zeng. Tracking by instance detection: A meta-learning approach. In *Proceedings of the IEEE/CVF Conference on Computer Vision and Pattern Recognition*, pages 6288–6297, 2020.
48. Yaqing Wang, Quanming Yao, James T Kwok, and Lionel M Ni. Generalizing from a few examples: A survey on few-shot learning. *ACM Computing Surveys (CSUR)*, 53(3):1–34, 2020.
49. Yu-Xiong Wang, Deva Ramanan, and Martial Hebert. Meta-learning to detect rare objects. In *Proceedings of the IEEE International Conference on Computer Vision*, pages 9925–9934, 2019.

50. Xiongwei Wu, Doyen Sahoo, and Steven CH Hoi. Meta-rcnn: Meta learning for few-shot object detection. 2019.
51. Ting Yao, Yingwei Pan, Yehao Li, Zhaofan Qiu, and Tao Mei. Boosting image captioning with attributes. In *Proceedings of the IEEE International Conference on Computer Vision*, pages 4894–4902, 2017.
52. Quanzeng You, Hailin Jin, Zhaowen Wang, Chen Fang, and Jiebo Luo. Image captioning with semantic attention. In *Proceedings of the IEEE conference on computer vision and pattern recognition*, pages 4651–4659, 2016.
53. Zheng Zhu, Qiang Wang, Bo Li, Wei Wu, Junjie Yan, and Weiming Hu. Distractor-aware siamese networks for visual object tracking. In *Proceedings of the European Conference on Computer Vision (ECCV)*, pages 101–117, 2018.
54. Alberti, C., Ling, J., Collins, M., Reitter, D.: Fusion of detected objects in text for visual question answering. arXiv preprint arXiv:1908.05054 (2019)
55. Andreas, J., Rohrbach, M., Darrell, T., Klein, D.: Neural module networks. In: Proceedings of the IEEE conference on computer vision and pattern recognition, pp. 39–48 (2016)
56. Chen, D., Dolan, W.B.: Collecting highly parallel data for paraphrase evaluation. In: Proceedings of the 49th Annual Meeting of the Association for Computational Linguistics: Human Language Technologies, pp. 190–200 (2011)
57. Chen, K., Wang, J., Chen, L.C., Gao, H., Xu, W., Nevatia, R.: Abc-cnn: An attention based convolutional neural network for visual question answering. arXiv preprint arXiv:1511.05960 (2015)
58. Chen, W., Gan, Z., Li, L., Cheng, Y., Wang, W., Liu, J.: Meta module network for compositional visual reasoning. In: Proceedings of the IEEE/CVF Winter Conference on Applications of Computer Vision, pp. 655–664 (2021)
59. Das, A., Kottur, S., Gupta, K., Singh, A., Yadav, D., Moura, J.M., Parikh, D., Batra, D.: Visual dialog. In: Proceedings of the IEEE Conference on Computer Vision and Pattern Recognition, pp. 326–335 (2017)
60. Fukui, A., Park, D.H., Yang, D., Rohrbach, A., Darrell, T., Rohrbach, M.: Multimodal compact bilinear pooling for visual question answering and visual grounding. arXiv preprint arXiv:1606.01847 (2016)
61. Gao, H., Mao, J., Zhou, J., Huang, Z., Wang, L., Xu, W.: Are you talking to a machine? dataset and methods for multilingual image question answering. arXiv preprint arXiv:1505.05612 (2015)
62. Girdhar, R., Ramanan, D.: Cater: A diagnostic dataset for compositional actions and temporal reasoning. arXiv preprint arXiv:1910.04744 (2019)
63. Goodfellow, I., Bengio, Y., Courville, A.: Deep Learning. MIT Press (2016). http://www.deeplearningbook.org
64. Goyal, Y., Khot, T., Summers-Stay, D., Batra, D., Parikh, D.: Making the v in vqa matter: Elevating the role of image understanding in visual question answering. In: Proceedings of the IEEE Conference on Computer Vision and Pattern Recognition, pp. 6904–6913 (2017)
65. Hudson, D.A., Manning, C.D.: Compositional attention networks for machine reasoning. arXiv preprint arXiv:1803.03067 (2018)
66. Hudson, D.A., Manning, C.D.: Gqa: A new dataset for real-world visual reasoning and compositional question answering. In: Proceedings of the IEEE/CVF Conference on Computer Vision and Pattern Recognition, pp. 6700–6709 (2019)
67. Jang, Y., Song, Y., Yu, Y., Kim, Y., Kim, G.: Tgif-qa: Toward spatio-temporal reasoning in visual question answering. In: Proceedings of the IEEE Conference on Computer Vision and Pattern Recognition, pp. 2758–2766 (2017)
68. Johnson, J., Hariharan, B., van der Maaten, L., Fei-Fei, L., Lawrence Zitnick, C., Girshick, R.: Clevr: A diagnostic dataset for compositional language and elementary visual reasoning. In: Proceedings of the IEEE Conference on Computer Vision and Pattern Recognition, pp. 2901–2910 (2017)

69. Johnson, J., Hariharan, B., Van Der Maaten, L., Hoffman, J., Fei-Fei, L., Lawrence Zitnick, C., Girshick, R.: Inferring and executing programs for visual reasoning. In: Proceedings of the IEEE International Conference on Computer Vision, pp. 2989–2998 (2017)
70. Kim, J.H., Lee, S.W., Kwak, D.H., Heo, M.O., Kim, J., Ha, J.W., Zhang, B.T.: Multimodal residual learning for visual qa. arXiv preprint arXiv:1606.01455 (2016)
71. Krishna, R., Zhu, Y., Groth, O., Johnson, J., Hata, K., Kravitz, J., Chen, S., Kalantidis, Y., Li, L.J., Shamma, D.A., et al.: Visual genome: Connecting language and vision using crowdsourced dense image annotations. International journal of computer vision **123**(1), 32–73 (2017)
72. Lei, J., Yu, L., Bansal, M., Berg, T.L.: Tvqa: Localized, compositional video question answering. arXiv preprint arXiv:1809.01696 (2018)
73. Lei, J., Yu, L., Berg, T.L., Bansal, M.: Tvqa+: Spatio-temporal grounding for video question answering. arXiv preprint arXiv:1904.11574 (2019)
74. Li, G., Duan, N., Fang, Y., Gong, M., Jiang, D.: Unicoder-vl: A universal encoder for vision and language by cross-modal pre-training. In: Proceedings of the AAAI Conference on Artificial Intelligence, vol. 34, pp. 11336–11344 (2020)
75. Li, L.H., Yatskar, M., Yin, D., Hsieh, C.J., Chang, K.W.: Visualbert: A simple and performant baseline for vision and language. arXiv preprint arXiv:1908.03557 (2019)
76. Li, Y., Song, Y., Cao, L., Tetreault, J., Goldberg, L., Jaimes, A., Luo, J.: Tgif: A new dataset and benchmark on animated gif description. In: Proceedings of the IEEE Conference on Computer Vision and Pattern Recognition, pp. 4641–4650 (2016)
77. Lin, T.Y., Maire, M., Belongie, S., Hays, J., Perona, P., Ramanan, D., Dollár, P., Zitnick, C.L.: Microsoft coco: Common objects in context. In: European conference on computer vision, pp. 740–755. Springer (2014)
78. Lu, J., Batra, D., Parikh, D., Lee, S.: Vilbert: Pretraining task-agnostic visiolinguistic representations for vision-and-language tasks. arXiv preprint arXiv:1908.02265 (2019)
79. Lu, J., Yang, J., Batra, D., Parikh, D.: Hierarchical question-image co-attention for visual question answering. arXiv preprint arXiv:1606.00061 (2016)
80. Ma, L., Lu, Z., Li, H.: Learning to answer questions from image using convolutional neural network. In: Proceedings of the AAAI Conference on Artificial Intelligence, vol. 30 (2016)
81. Malinowski, M., Rohrbach, M., Fritz, M.: Ask your neurons: A neural-based approach to answering questions about images. In: Proceedings of the IEEE international conference on computer vision, pp. 1–9 (2015)
82. Mao, J., Gan, C., Kohli, P., Tenenbaum, J.B., Wu, J.: The neuro-symbolic concept learner: Interpreting scenes, words, and sentences from natural supervision. arXiv preprint arXiv:1904.12584 (2019)
83. Noh, H., Seo, P.H., Han, B.: Image question answering using convolutional neural network with dynamic parameter prediction. In: Proceedings of the IEEE conference on computer vision and pattern recognition, pp. 30–38 (2016)
84. Ren, M., Kiros, R., Zemel, R.: Image question answering: A visual semantic embedding model and a new dataset. Proc. Advances in Neural Inf. Process. Syst **1**(2), 5 (2015)
85. Saito, K., Shin, A., Ushiku, Y., Harada, T.: Dualnet: Domain-invariant network for visual question answering. In: 2017 IEEE International Conference on Multimedia and Expo (ICME), pp. 829–834. IEEE (2017)
86. Singh, A., Natarajan, V., Shah, M., Jiang, Y., Chen, X., Batra, D., Parikh, D., Rohrbach, M.: Towards vqa models that can read. In: Proceedings of the IEEE/CVF Conference on Computer Vision and Pattern Recognition, pp. 8317–8326 (2019)
87. Su, W., Zhu, X., Cao, Y., Li, B., Lu, L., Wei, F., Dai, J.: Vl-bert: Pre-training of generic visual-linguistic representations. arXiv preprint arXiv:1908.08530 (2019)
88. Sun, C., Myers, A., Vondrick, C., Murphy, K., Schmid, C.: Videobert: A joint model for video and language representation learning. In: Proceedings of the IEEE/CVF International Conference on Computer Vision, pp. 7464–7473 (2019)
89. Tan, H., Bansal, M.: Lxmert: Learning cross-modality encoder representations from transformers. arXiv preprint arXiv:1908.07490 (2019)

90. Teney, D., van den Hengel, A.: Visual question answering as a meta learning task. In: Proceedings of the European Conference on Computer Vision (ECCV), pp. 219–235 (2018)
91. Wang, P., Wu, Q., Shen, C., Hengel, A.v.d., Dick, A.: Explicit knowledge-based reasoning for visual question answering. arXiv preprint arXiv:1511.02570 (2015)
92. Wu, Q., Teney, D., Wang, P., Shen, C., Dick, A., van den Hengel, A.: Visual question answering: A survey of methods and datasets. Computer Vision and Image Understanding **163**, 21–40 (2017)
93. Wu, Q., Wang, P., Shen, C., Dick, A., Van Den Hengel, A.: Ask me anything: Free-form visual question answering based on knowledge from external sources. In: Proceedings of the IEEE conference on computer vision and pattern recognition, pp. 4622–4630 (2016)
94. Xiong, C., Merity, S., Socher, R.: Dynamic memory networks for visual and textual question answering. In: International conference on machine learning, pp. 2397–2406. PMLR (2016)
95. Xu, H., Saenko, K.: Ask, attend and answer: Exploring question-guided spatial attention for visual question answering. In: European Conference on Computer Vision, pp. 451–466. Springer (2016)
96. Xu, J., Mei, T., Yao, T., Rui, Y.: Msr-vtt: A large video description dataset for bridging video and language. In: Proceedings of the IEEE conference on computer vision and pattern recognition, pp. 5288–5296 (2016)
97. Yang, Z., He, X., Gao, J., Deng, L., Smola, A.: Stacked attention networks for image question answering. In: Proceedings of the IEEE conference on computer vision and pattern recognition, pp. 21–29 (2016)
98. Yi, K., Gan, C., Li, Y., Kohli, P., Wu, J., Torralba, A., Tenenbaum, J.B.: Clevrer: Collision events for video representation and reasoning. arXiv preprint arXiv:1910.01442 (2019)
99. Yi, K., Wu, J., Gan, C., Torralba, A., Kohli, P., Tenenbaum, J.B.: Neural-symbolic vqa: Disentangling reasoning from vision and language understanding. arXiv preprint arXiv:1810.02338 (2018)
100. Zellers, R., Bisk, Y., Farhadi, A., Choi, Y.: From recognition to cognition: Visual commonsense reasoning. In: Proceedings of the IEEE/CVF Conference on Computer Vision and Pattern Recognition, pp. 6720–6731 (2019)
101. Zhu, Y., Groth, O., Bernstein, M., Fei-Fei, L.: Visual7w: Grounded question answering in images. In: Proceedings of the IEEE conference on computer vision and pattern recognition, pp. 4995–5004 (2016)
102. Blitzer, J., Dredze, M., Pereira, F.: Biographies, bollywood, boom-boxes and blenders: Domain adaptation for sentiment classification. In: Proceedings of the 45th annual meeting of the association of computational linguistics, pp. 440–447 (2007)
103. Chen, M., Zhang, W., Zhang, W., Chen, Q., Chen, H.: Meta relational learning for few-shot link prediction in knowledge graphs. arXiv preprint arXiv:1909.01515 (2019)
104. Coucke, A., Saade, A., Ball, A., Bluche, T., Caulier, A., Leroy, D., Doumouro, C., Gisselbrecht, T., Caltagirone, F., Lavril, T., et al.: Snips voice platform: an embedded spoken language understanding system for private-by-design voice interfaces. arXiv preprint arXiv:1805.10190 (2018)
105. Deng, S., Zhang, N., Kang, J., Zhang, Y., Zhang, W., Chen, H.: Meta-learning with dynamic-memory-based prototypical network for few-shot event detection. In: Proceedings of the 13th International Conference on Web Search and Data Mining, pp. 151–159 (2020)
106. Finn, C., Abbeel, P., Levine, S.: Model-agnostic meta-learning for fast adaptation of deep networks. In: International Conference on Machine Learning, pp. 1126–1135. PMLR (2017)
107. Gao, T., Han, X., Liu, Z., Sun, M.: Hybrid attention-based prototypical networks for noisy few-shot relation classification. In: Proceedings of the AAAI Conference on Artificial Intelligence, vol. 33, pp. 6407–6414 (2019)
108. Geng, R., Li, B., Li, Y., Zhu, X., Jian, P., Sun, J.: Induction networks for few-shot text classification. arXiv preprint arXiv:1902.10482 (2019)
109. Gu, J., Wang, Y., Chen, Y., Cho, K., Li, V.O.: Meta-learning for low-resource neural machine translation. arXiv preprint arXiv:1808.08437 (2018)

110. Han, X., Zhu, H., Yu, P., Wang, Z., Yao, Y., Liu, Z., Sun, M.: Fewrel: A large-scale supervised few-shot relation classification dataset with state-of-the-art evaluation. arXiv preprint arXiv:1810.10147 (2018)
111. Holla, N., Mishra, P., Yannakoudakis, H., Shutova, E.: Learning to learn to disambiguate: Meta-learning for few-shot word sense disambiguation. arXiv preprint arXiv:2004.14355 (2020)
112. Jiang, X., Havaei, M., Chartrand, G., Chouaib, H., Vincent, T., Jesson, A., Chapados, N., Matwin, S.: Attentive task-agnostic meta-learning for few-shot text classification (2018)
113. Lai, V.D., Dernoncourt, F., Nguyen, T.H.: Exploiting the matching information in the support set for few shot event classification. In: Pacific-Asia Conference on Knowledge Discovery and Data Mining, pp. 233–245. Springer (2020)
114. Larson, S., Mahendran, A., Peper, J.J., Clarke, C., Lee, A., Hill, P., Kummerfeld, J.K., Leach, K., Laurenzano, M.A., Tang, L., et al.: An evaluation dataset for intent classification and out-of-scope prediction. arXiv preprint arXiv:1909.02027 (2019)
115. Lv, X., Gu, Y., Han, X., Hou, L., Li, J., Liu, Z.: Adapting meta knowledge graph information for multi-hop reasoning over few-shot relations. arXiv preprint arXiv:1908.11513 (2019)
116. Madotto, A., Lin, Z., Wu, C.S., Fung, P.: Personalizing dialogue agents via meta-learning. In: Proceedings of the 57th Annual Meeting of the Association for Computational Linguistics, pp. 5454–5459 (2019)
117. Mi, F., Huang, M., Zhang, J., Faltings, B.: Meta-learning for low-resource natural language generation in task-oriented dialogue systems. arXiv preprint arXiv:1905.05644 (2019)
118. Obamuyide, A., Vlachos, A.: Model-agnostic meta-learning for relation classification with limited supervision (2020)
119. Qian, K., Yu, Z.: Domain adaptive dialog generation via meta learning. arXiv preprint arXiv:1906.03520 (2019)
120. Sui, D., Chen, Y., Mao, B., Qiu, D., Liu, K., Zhao, J.: Knowledge guided metric learning for few-shot text classification. arXiv preprint arXiv:2004.01907 (2020)
121. Triantafillou, E., Zhu, T., Dumoulin, V., Lamblin, P., Evci, U., Xu, K., Goroshin, R., Gelada, C., Swersky, K., Manzagol, P.A., et al.: Meta-dataset: A dataset of datasets for learning to learn from few examples. arXiv preprint arXiv:1903.03096 (2019)
122. Xia, C., Zhang, C., Nguyen, H., Zhang, J., Yu, P.: Cg-bert: Conditional text generation with bert for generalized few-shot intent detection. arXiv preprint arXiv:2004.01881 (2020)
123. Xia, C., Zhang, C., Yan, X., Chang, Y., Yu, P.S.: Zero-shot user intent detection via capsule neural networks. arXiv preprint arXiv:1809.00385 (2018)
124. Yin, W.: Meta-learning for few-shot natural language processing: A survey. arXiv preprint arXiv:2007.09604 (2020)
125. Yu, M., Guo, X., Yi, J., Chang, S., Potdar, S., Cheng, Y., Tesauro, G., Wang, H., Zhou, B.: Diverse few-shot text classification with multiple metrics. arXiv preprint arXiv:1805.07513 (2018)

Chapter 5
Future Research Directions

In this chapter, we provide the readers with several potential future research directions on AutoML (hyper-parameter optimization and neural architecture search) and meta-learning. We would like to point out that the future works deserving further investigations discussed in this chapter can also be applicable to various multimedia applications as well as the learning tasks of different modalities (i.e., texts, audios, images, and videos), therefore we do not present separate discussions here.

5.1 On Hyper-Parameter Optimization

We point out two research directions that should draw more attentions from the community: evaluation metric and the ability of generalization.

Given the vast amount of HPO methods developed based on diverse methodologies, one may naturally ask what are the strengths and weaknesses of each method and, more practically, how to evaluate the fitness of different methods to specific problem instances. Indeed, to promote fairness in comparison, we may desire to use a common benchmark for evaluation. One criterion specifically needed is the metric to judge the effectiveness of an HPO method. However, most current works in HPO adopt different performance metrics, making it difficult to give an overall comparison for general methods. For example, whether the performance should be evaluated on the validation set or the test set, as well as after a fixed number of function evaluations or after a given amount of time. Such differences will surely affect the objective as well as the final judgment, and thus should be handled with thorough care when deciding the method for a specific task instance.

Moreover, overfitting is another open problem of HPO. Recall that, in its problem formulation, the test performance is estimated by the validation performance, which is only evaluated on a finite number of instance points. Therefore, the hyperparameters are prone to overfit to the validation set, and they do not necessarily work

W. Zhu, X. Wang, *Automated Machine Learning and Meta-Learning for Multimedia*,
https://doi.org/10.1007/978-3-030-88132-0_5

equally well on the real task instance. Recently many heuristic approaches have been proposed to resolve the overfitting issue, such as applying a different train-validation split to each function evaluation or using a separate holdout set to specifically assess configurations produced by HPO. However, despite all these different approaches, there is no commonly agreed technique to best handle overfitting, nor theoretical analysis on the generalization ability of current AutoML methods.

5.2   On Neural Architecture Search

The emergence of neural architecture search (NAS) is an exciting development, which aims to end the tedious trial and error process of manual neural architecture design. Current NAS methods have achieved better or comparable performances in a variety of domains. However, NAS is still at its early stages since lots of problems have not been solved or explored in the research fields. When applying NAS to multimedia scenarios, NAS helps to automatically design effective neural networks for multimedia search and recommendation, as well as a wide range of tasks of different modalities (especially on computer vision and natural language processing tasks) and multi-modal applications, as is introduced in Chap. 3. The following discussion on future directions of NAS is expected to benefit its multimedia applications in the coming years.

The original expectation of NAS is to design neural architecture without manual participation. Thus, in the early stage of NAS, global search space is used to give enough freedom for NAS algorithms to design a new architecture. But searching in such a large space costs lots of computational resources, NAS methods need a very long time to explore the space. The huge cost limits the chances for researchers to develop NAS fields. Subsequent works begin to use cell-based search space, which makes it possible for researchers to run NAS in a relatively short time while getting an outstanding performance. Nevertheless, designing a cell-based search space already needs lots of human's domain knowledge to decide the restrictions inside cells and how cells construct the whole architecture, which limits the freedom for NAS methods to design novel architectures, departures from the original purpose of NAS. Besides, architectures searched in smaller spaces often outperform those in larger spaces, making it unclear why those better architectures are missed during searching in larger spaces. All those results show that current NAS methods are not enough to explore a large architecture space and find a perfect architecture yet. How to efficiently explore the architecture space is still the core question in NAS.

One direction to help explore a large search space may be the pre-training of NAS models. Pre-trained models have been widely used in the natural language processing domain. To efficiently get accurate word embeddings, researchers can directly use pre-train models which have been trained on a huge corpus. When faced with a new task, only a light fine-tuning on the pre-train model is needed since it has already learned lots of useful knowledge from other datasets. This technique can also be used in NAS. Since the tasks within the same domain may need similar architectures, NAS methods can directly borrow knowledge from similar datasets. Some NAS works have developed works that combine transfer learning and meta-learning in NAS, aiming to design architectures with knowledge of other datasets. Pre-train models may be the next step of works in this direction.

Another issue that draws lots of attention is robustness. Along with other domains of machine learning, especially computer vision and graph representation learning, NAS meets the risk of adversarial attacks as well. When NAS has been widely used in real-world applications, attackers may adopt perturbation on the data to fool NAS methods, making it generate bad architectures and causing crashes. However, this issue has not been studied. There are two types of robustness issues in the NAS scheme. One is on the architecture search level—it requires NAS algorithms to have the ability of defending adversarial attacks and finding good architectures fitting the correct data distributions. The other one is on architecture space level—it requires NAS algorithm to have the chances of designing architectures possessing the ability of defendance. The latter one may be easier to achieve if considering current works on robustness issues on machine learning domains. But the former one is riskier and needs effort on this issue.

NAS also lacks interpretability currently. Classical NAS methods such as RL and EA treat the NAS problem as a black box, not caring about why the designed architecture is good. One-shot NAS methods use weights or scores to indicate the importance of operation candidates, which cannot interpret the designing process as well. If NAS methods can interpret the reason why an operation over the input source should be chosen at each layer, it will provide human with precious knowledge about how deep neural networks work and with inspirations to design more useful architectures. Current works have found some regular patterns of how NAS methods design architectures, but we still need more effort in this direction.

In a nutshell, the development of NAS excites researchers in the past few years. Nevertheless, NAS is still at the beginning stage, and additional theoretical guidance and experimental analysis are required. There is still a long way to go before we are able to totally replace manually designed architectures with NAS.

5.3   On Meta-Learning

Traditional machine learning models are trained offline with the given complete dataset and tested once without updating models. In practice, it is more general that the given dataset arrives sequentially with dynamic drifts of distribution. The machine learning models should consolidate knowledge learned from ever-seen data while learning new knowledge from newly coming data. Besides, the knowledge can be transferred across the data stream. Considering these more general machine learning scenarios with streaming datasets, online meta-learning and continual learning attract an increasing number of interests recently. Investigating the online meta-learning and continual learning problems benefits a large number of real-world applications, such as robot imitation learning, autonomous driving, etc.

As we have discussed in this book, meta-learning is widely applied to search and recommendation, classification, detection, recognition, tracking, question answering etc. In real-world scenarios, the datasets of these applications are also usually organized as a form of the data stream. Therefore, the following discussion on online meta-learning and continual learning provides promising future directions for applying meta-learning to multimedia.

Online meta-learning focuses on handling non-stationary task distribution with various tasks arriving in sequence, while traditional meta-learning assumes that tasks are sampled from the same stationary distribution. The goal of an online meta-learner is to facilitate the new task with the learned knowledge of previous tasks. The key challenge is that the distribution of tasks varies over time. Several existing works [12, 13] tackle the issue by minimizing the regret of online meta-learning and other work [4] captures the changing of tasks using the Dirichlet process mixture model. For future investigations, the relationships among streaming tasks could be further explored and the implicit structure of task distribution could be analyzed theoretically.

The forward transfer of knowledge from old tasks to new tasks is studied in online meta-learning. Furthermore, continual learning studies both the forward transfer and catastrophic forgetting problems. Faced with sequential tasks, an agent in continual learning should learn the new task as well as consolidate the memory of old tasks. In human learning, we can continuously learn knowledge from the real world without overwriting previous old knowledge. Therefore, how to learn continuously with new tasks arriving and consolidate the knowledge obtained from old tasks is the core challenge in continual learning. Numerous methods for alleviating catastrophic forgetting have been proposed. Expansion-based approaches [7, 14] isolate model parameters to achieve the purpose of continuously learning new knowledge without forgetting the old knowledge. Prior-based approaches [1, 3, 6, 15] make assumptions on the distribution of parameters and force them to drift away from the feasible regions of previous tasks. Replay-based methods [2, 5, 8–11] store data samples or generate rehearsal samples of previous tasks to address forgetting issue. Most existing approaches rely on the explicit indications of task switches. However, the distribution of tasks may change without indication and even shift gradually in an online scenario. In the future, the study of online continual learning is a promising area. Besides, unsupervised continual learning can be further studied since there are a large number of unlabeled data in the real world.

References

1. Arslan Chaudhry, Puneet Kumar Dokania, Thalaiyasingam Ajanthan, and Philip H S Torr. Riemannian walk for incremental learning: Understanding forgetting and intransigence. *european conference on computer vision*, pages 556–572, 2018.

2. Arslan Chaudhry, Marcus Rohrbach, Mohamed Elhoseiny, Thalaiyasingam Ajanthan, Puneet Kumar Dokania, Philip H S Torr, and Marcaurelio Ranzato. On tiny episodic memories in continual learning. *arXiv: Learning*, 2019.
3. Ferenc Huszar. Note on the quadratic penalties in elastic weight consolidation. *Proceedings of the National Academy of Sciences of the United States of America*, 115(11), 2018.
4. Ghassen Jerfel, Erin Grant, Thomas L Griffiths, and Katherine Heller.   Reconciling meta-learning and continual learning with online mixtures of tasks.   *arXiv preprint arXiv:1812.06080*, 2018.
5. Nitin Kamra, Umang Gupta, and Yan Liu. Deep generative dual memory network for continual learning. *arXiv: Learning*, 2018.
6. James Kirkpatrick, Razvan Pascanu, Neil C Rabinowitz, Joel Veness, Guillaume Desjardins, Andrei A Rusu, Kieran Milan, John Quan, Tiago Ramalho, Agnieszka Grabskabarwinska, et al. Overcoming catastrophic forgetting in neural networks. *Proceedings of the National Academy of Sciences of the United States of America*, 114(13):3521–3526, 2017.
7. Xilai Li, Yingbo Zhou, Tianfu Wu, Richard Socher, and Caiming Xiong. Learn to grow: A continual structure learning framework for overcoming catastrophic forgetting. *arXiv preprint arXiv:1904.00310*, 2019.
8. Zhizhong Li and Derek Hoiem. Learning without forgetting. *european conference on computer vision*, 40(12):614–629, 2018.
9. David Lopez-Paz and Marc'Aurelio Ranzato.   Gradient episodic memory for continual learning. In *Advances in Neural Information Processing Systems*, pages 6467–6476, 2017.
10. Sylvestrealvise Rebuffi, Alexander Kolesnikov, Georg Sperl, and Christoph H Lampert. icarl: Incremental classifier and representation learning. pages 5533–5542, 2017.
11. Hanul Shin, Jung Kwon Lee, Jaehong Kim, and Ji Won Kim. Continual learning with deep generative replay. *arXiv: Artificial Intelligence*, 2017.
12. Huaxiu Yao, Ying Wei, Junzhou Huang, and Zhenhui Li.   Hierarchically structured meta-learning. In *International Conference on Machine Learning*, pages 7045–7054. PMLR, 2019.
13. Huaxiu Yao, Yingbo Zhou, Mehrdad Mahdavi, Zhenhui Li, Richard Socher, and Caiming Xiong. Online structured meta-learning. *arXiv preprint arXiv:2010.11545*, 2020.
14. Jaehong Yoon, Eunho Yang, Jeongtae Lee, and Sung Ju Hwang.   Lifelong learning with dynamically expandable networks. *arXiv: Learning*, 2017.
15. Friedemann Zenke, Ben Poole, and Surya Ganguli.   Continual learning through synaptic intelligence. *arXiv: Learning*, 2017.

Index

Printed in the United States
by Baker & Taylor Publisher Services